Cláudio Silva Soares
Riselane de L A Bruno
José Pires Dantas

Rhizobium Efficiency in Cowpeas (Vigna unguiculata L. Walp.)

Cláudio Silva Soares
Riselane de L A Bruno
José Pires Dantas

Rhizobium Efficiency in Cowpeas (Vigna unguiculata L. Walp.)

Yield and Physiological Quality of Seeds

ScienciaScripts

Imprint

Cover image: www.ingimage.com

This book is a translation from the original published under ISBN 978-3-330-75681-6.

Publisher:
Sciencia Scripts
is a trademark of
Dodo Books Indian Ocean Ltd. and OmniScriptum S.R.L publishing group

120 High Road, East Finchley, London, N2 9ED, United Kingdom
Str. Armeneasca 28/1, office 1, Chisinau MD-2012, Republic of Moldova, Europe
Printed at: see last page
ISBN: 978-620-8-28134-2

SUMMARY

To all those who have always believed in the realization of this ideal.

I dedicate

Especially to my parents, children and wife.

I offer

CHAPTER I. COWPEA, BIOLOGICAL NITROGEN FIXATION AND NITROGEN FERTILIZATION

1.1. INTRODUCTION

The northeast of Brazil has a vast diversity of species cultivated by small and medium-sized farmers and, among these, the macassar bean or cowpea [*Vigna unguiculata* (L.) Walp.] stands out for its consumption preference in this region. Despite its socio-economic importance, the majority of this crop is still grown precariously and without the use of appropriate technologies that promote better yields through sustainable agricultural management alternatives.

Duarte (1998), discussing agricultural development in the Cerrados, points out that in Brazil, globalization and the modernization of agriculture have brought with them the degradation and depletion of natural resources. The five most important agricultural products that contribute to exports are produced using agricultural practices that are, in general, highly sensitive to the relationship between the environment, agriculture and development (rural and national), as they are in one way or another associated with deforestation, erosion and the contamination of soils and water sources.

According to Assad & Almeida (2004), from the point of view of basic technologies for sustainable agriculture, there are often two types of obstacle. The first concerns the technologies themselves which, although sometimes known and tested on a scientific basis, are not properly inserted into production systems, either because of a lack of appropriate technological dissemination or because of a lack of coordination between research and rural extension and the productive segments that could benefit from these technologies. Another obstacle is the more or less widespread difficulty in deepening knowledge about agricultural systems or the lack of clarity about their dynamics.

The use of a rhizobial strain, recognized as capable of increasing the productivity of cowpeas in rainfed areas, represents an opportunity for small and medium-sized farmers in the semi-arid region to increase the supply of grain and make extra profits. To this end, the commercialization of this strain by inoculant industries depends on its official recommendation according to the results of agronomic efficiency obtained in line with the protocol established by RELARE (Network of laboratories for the recommendation, standardization and dissemination of technology for microbial inoculants of agricultural interest).

Although there are already three strains recommended for cowpeas, one of them obtained from previous partnerships between members of this project, there is a wide variability in the response of these strains in the field, which may be related to cultivar specificity.

In this context, the aim of this study was to evaluate rhizobium strains for inoculation in the cowpea cultivar "Corujinha", in order to provide small and medium-sized farmers with alternatives that can increase their income and at the same time conserve the natural resources on their property.

1.2. Characterization and importance of cowpeas

The cowpea [*Vigna unguiculata* (L.) Walp] is a *Fabaceae* herbaceous plant, native to Africa and widely cultivated in the tropical regions of the African, Asian and American continents (Soares et al., 2006a). According to Oliveira et al. (2002), it is known in the Northeast of Brazil as macassar bean or string bean, and is considered a source of alternative income and staple food for its population, as it is one of the main crops in this region.

Its nutritional properties, which are relatively superior to those of common beans, and the low cost of production mean that this crop is considered extremely important in social and economic terms for the region (Araùjo & Watt, 1988).

The composition of cowpeas varies depending on the cultivar, as Castellón et al, (2003), analyzing six cowpea cultivars (Br 14, Br 9, Br 17, CNC 0434, Vita 7 and CE 315), found these changes in crude protein (21.6 to 24.7%), total lipids (1.4 to 1.7%), ash (2.3 to 3.2%) and carbohydrates (73.5 to 74.5).

According to Vieira et al. (2000), this crop can be eaten in the form of green pods, green kernels and dried kernels, as well as in other forms of preparation, such as acarajé, for example. In the form of green pods, the crop is harvested when the pods are well developed but have little fiber. For consumption in the form of green beans, the pods are harvested when they begin to ripen. For dry grain production, the pods are harvested when they are completely dry.

In addition, its stems and branches are usually used in animal feed (Silva & Oliveira, 1993), or even, due to its hardiness and ability to grow satisfactorily in low-fertility soils, this crop is considered an excellent option as a source of organic matter for recovering soils that are naturally poor in fertility, or depleted by intensive use, which is very common in the Northeast.

The most suitable temperature for the development of cowpeas is between 20 and 35 °C (Araùjo et al., 1984), which is higher than that suitable for common beans (Andrade, 1998).

Therefore, the cowpea is a crop best suited to regions where rainfall is very irregular and scarce, especially in the municipalities of the semi-arid northeast.

Soares et al. (2006) point out that, although little known in central-southern Brazil, the cowpea is of great socio-economic importance and has notable strategic potential, especially for the North and Northeast regions, where it is one of the most important components of the diet. According to these authors, the traditional production structure and the restricted market have been changing every year, and it is now also grown by medium and large producers, with greater adoption of technology. It is already being marketed beyond regional borders, as is the case with black-eyed peas on commodities exchanges in the Southeast, and has broad export prospects, since there is an international market for types traditionally grown in Brazil.

Despite being considered a tropical crop, compatible with local ecological conditions, it still has low productivity, both in the single and intercropped systems (Miranda et al., 1996). Among the main causes limiting the productivity of cowpeas in the Northeast, Freire Filho et al. (1998) highlight the use of traditional cultivars with low productive capacity, around 400 to 500 kg ha $.^{-1}$

1.3. Biological nitrogen fixation

Biological nitrogen fixation (BNF) is characterized by the conversion of gaseous nitrogen (N2) into ammoniacal nitrogen (NH4), a form available to plants. According to Taiz & Zeiger (2004), the process of biological nitrogen fixation is similar to industrial nitrogen fixation, as it produces ammonia from molecular nitrogen, the general reaction being as follows: N2 + 8^{e-} + 8 H + 16 ATP $\rightarrow$ 2 NH_3 + H_2 + 16 ADP + 16 Pi.

Nitrogen-fixing species have a complex of enzymes (nitrogenase complex) responsible for catalyzing this reaction. Motta (2003) reports that the structure of this enzyme consists of two proteins called dinitrogenase (MoFe-protein) and dinitrogenase-reductase (Fé-protein).

This process occurs as a result of the symbiotic association between leguminous plants and specific bacteria that associate with their roots, forming nodules that serve to accumulate this nutrient. Once the symbiosis has been established, the plant provides photoassimilates to the bacteria, receiving nitrogenous products (amino acids, ureides) from N2 fixation in return (Schubert, 1986).

Nitrogen is present in various forms in the biosphere. The atmosphere contains a vast amount (around 78% by volume) of molecular nitrogen (N2). However, this large reservoir of nitrogen is not directly available to living organisms. Most plants obtain nitrogen from the soil in the form of nitrate ion (NO_3^-), while some absorb it in the form of ammonium ion (NH_4^+). Obtaining nitrogen from the atmosphere requires breaking a triple covalent bond of exceptional stability between the two nitrogen atoms (N≡N) to produce ammonia (NH_3) or nitrate (NO_3^-). These reactions, known as nitrogen fixation, can be obtained by industrial and natural processes (Taiz & Zeiger, 2004).

In nitrate assimilation (NO_3^-), the nitrogen in NO_3^- is converted into a more energetic form, nitrite (NO_2^-), which in turn is transformed into an even more energetic form, ammonium (NH_4^+), and finally into glutamine amide nitrogen. This process consumes the equivalent of 12 ATPs for each nitrogen (Bloom et al., 1992).

Natural processes fix around 190 x 10^{12} g $year^{-1}$ of nitrogen through the following processes (Schlesinger, 1997):

• Lightning - Lightning is responsible for approximately 8% of the nitrogen fixed. They convert water vapor and oxygen into highly reactive free hydroxyl radicals, free hydrogen atoms and free oxygen atoms, which attack molecular nitrogen (N_2) to form nitric acid (HNO_3). This nitric acid then precipitates on Earth as rain.

• Photochemical reactions - Around 2% of the nitrogen fixed is derived from photochemical reactions between gaseous nitric oxide (NO) and ozone (O_3), producing nitric acid (HNO_3).

• Biological nitrogen fixation - 90% of the remaining nitrogen results from biological fixation, in which bacteria or blue algae (cyanobacteria)

fix N2 into ammonium (NH_4^+).

These bacteria belong to the *Rhizobiaceae* family and are associated with the roots of various legumes, including cowpeas, playing a fundamental role, which is to supply the plant's need for nitrogen with bacteria of the genus *Allorhizobium*, *Azorhizobium*, *Bradyrhizobium*, *Rhizobium*, *Sinorhizobium* and *Mesorhizobium*, which are generically called rhizobia (Vargas & Hungria 1997, Brandâo Jùnior & Hungria 2000).

There are several advantages to this process, ranging from increasing plant production to contributing to the sustainability of agricultural systems, recovering degraded areas and increasing soil fertility and organic matter. However, its main short-term advantage is associated with savings in the use of industrialized nitrogen fertilizers (Rumjanek et al., 2005).

Biological nitrogen fixation is mediated by a wide range of prokaryotic microorganisms with substantial morphological, physiological, genetic, biochemical and phylogenetic diversity.

This diversity guarantees the occurrence of this process in the most diverse terrestrial habitats. However, despite their great importance in maintaining the biosphere, it is estimated that less than 1% of the microorganisms on the planet have been characterized and described (Moreira & Siqueira 2002).

On the other hand, the great variability observed in the ability of strains to infect, nodulate and fix atmospheric nitrogen, associated with the influence exerted by the plant through intrinsic characteristics, can lead strains that are efficient at fixing N2 in some host legume species to show low fixing efficiency when associated with others (Xavier et al., 2006).

For the use of these species to be successful, it is necessary to use rhizobium strains that are efficient at fixing N2, taking into account, among other factors, competition with native populations established in the soil and adaptation to local environmental conditions, making it also necessary to study the density and diversity of the native rhizobia population (Soares et al., 2006).

The various environmental factors that can negatively affect the nodulation capacity and N fixation$_2$ of indigenous *Rhizobium* populations in soils include: pH, humidity, temperature, biological factors, salinity, nitrates, pesticides, herbicides and heavy metals (Castro, 2000).

Another relevant factor in the successful occupation of nodules by the bacteria is the symbiotic selectivity of plant genotypes, whose specificity may be the main factor in modifying *Rhizobium* populations in the soil (Paffetti *et al.*, 1998).

The selection of these strains is a stage of fundamental importance in the production of commercial inoculants for inoculation with efficient strains, seeking the best plant development in the field (Jesus et al., 2005).

In this sense, the existence of interactions between *Rhizobium meliloti* strains and alfalfa cultivars has been observed, with different results in terms of the efficiency of atmospheric nitrogen fixation (Vance et al., 1988; Arrese-Igor et al., 1989; Twary & Heichel, 1991). In Brazil, Kolling et al. (1983) tested the strains SEMIA-108 (Nitragin inoculant, USA); SEMIA-115 (USDA 1082, USA); SEMIA-116 (USDA 1088, USA); SEMIA-113 (SU 47, Australia); SEMIA-134 (isolate RS, Brazil) and SEMIA-135 (isolate RS, Brazil) in relation to the Crioula cultivar, and observed that strains SEMIA-115, SEMIA-116, SEMIA-113, SEMIA-134 and SEMIA-135 showed high efficiency in dry matter yield, while strain SEMIA-108 was the least efficient in the first year. In subsequent years, SEMIA-134 was less efficient than the other high-efficiency strains in the first year.

In the case of the soybean crop [*Glicine max* (L.) Merril], the best ability to compete for nodulation sites in the field has been attributed to the species *B. elkanii* (Boddey & Hungria, 1997; Neves & Rumjanek, 1997). However, in terms of N accumulation, *B. japonicum* has been found to be more efficient (Teaney & Fuhrmann, 1992; Neves & Rumjanek, 1997).

For the common bean (*Phaseolus vulgaris* L.), one of the main pieces of evidence for the different behavior of bean cultivars in relation to parameters related to BNF comes from studies carried out at the International Center for Tropical Agriculture (CIAT), where it was found that late maturing cultivars tended to accumulate more symbiotically fixed N (International Center for Tropical Agriculture, 1975).

From the available data on the morphological and molecular characteristics of beans, it has been found that there is a gene pool of Mesoamerican origin and another of Andean origin (Singh et al., 1991). Taking this observation into account, Kipe-Nolt et al. (1992) found a tendency for wild and domesticated materials of Mesoamerican origin to form nodules more quickly with *Rhizobium etli* strain CIAT 632 than with *R. tropici* CIAT 899 and materials from the Andes of Argentina and Peru to form nodules more quickly with CIAT 899.

Andriolo et al. (1994) compared domesticated and wild beans of different origins and found that the wild ones had a higher nodulation capacity.

Franco et al. (2002) also studied nodulation in bean cultivars from the Andean (WAF 15, WAF 7, WAF 6, Antioquia 8, Mineiro Precoce, Diacol Andino and Batatinha) and Mesoamerican (Ouro Negro, BAT 304, DOR 241, RAB 94, Ouro, FT 84835 and Ricopardo) and found that only a few cultivars (WAF 15, WAF 7, Mineiro Precoce, WAF 6 and Antioquia 8) showed specificity in nodulation. Considering all the cultivars and rhizobium strains, WAF 15 was the cultivar with the best performance in terms of nodule number and mass. WAF 6 was the worst performing cultivar, almost reaching the level of nodulation restriction with the *R. etli* CIAT 632 strain.

Ceccatto et al. (1998) observed a differential effect of bean cultivars from different gene pools under interaction with *R. tropici* on the activity of nodulins: leghemoglobin, phosphoenolpyruvate carboxylase and glutamine synthetase, as well as on nitrogenase activity and allantoin concentration.

With regard to cowpea [Vigna unguiculata (L.) Walp.], the association of this crop with rhizobium strains shows low specificity (Zilli et al., 2006). There are reports of association with at least six species of rhizobium: *Bradyrhizobium japonicum* (Jordan, 1982), *B. elkanii* (Kuykendall et al., 1992), *Sinorhizobium fredii* (Lajudie et al., 1994), *S. xinjiangensis* (Chen et al., 1988), *Rhizobium hainanense* (Chen et al., 1997) and *R. tropici* IIA (Zilli, 2001), as well as other *Bradyrhizobium* strains whose species have not been identified (Martins, 1996).

It is estimated that the average contribution of biological nitrogen fixation to cowpeas is between 73 and 240 kg ha^{-1} year .$^{-1}$

Studies of BNF in cowpea have focused mainly on the species *Bradyrhizobium japonicum* and *B. elkanii*, as they make the greatest contributions to BNF in most herbaceous legumes in tropical regions (Moreira & Siqueira, 2002).

With the cultivar BR-14 Mulato, it was found that inoculation in the field with the strains UFLA 03-84 (BR 3302) and INPA 03-11B (BR 3301) contributed significantly to the increase in grain yield, being similar to the control, with 70 kg ha^{-1} of N, and superior to the reference strain BR 2001 for this crop (Soares et al., 2006).

Other researchers (Zilli et al., 2006) also analyzed the symbiotic efficiency of some *Bradyrhizobium* strains isolated from soils in the Brazilian Cerrado and found that strains BR 3262, BR 3280 (characterized as *B. elkanii*) and BR 3267, BR 3287 and BR 3288 (*Bradyrhizobium* sp.) are potential inoculants for cowpeas, due to their good performance in both symbiotic efficiency and nodule occupancy.

Different strains also showed different responses when inoculating cowpea accessions from different nationalities, with Brazilian accessions showing the highest nodule occupancy rates, followed by those from Nigeria and the United States. The highest percentage of nodule occupancy in 6 of the 10 cowpea accessions tested was due to inoculation with strain BR 3273, and the lowest was due to strain BR 3269 in 8 of the 10 accessions. According to the authors, these data suggest a specificity between these strains and the cowpea accessions (Xavier et al., 2006).

Due to the existence of a great diversity of native species of nitrogen-fixing bacteria, which do so with a low degree of efficiency, it is necessary to obtain high-quality rhizobium strains capable of surviving and competing for the efficient fixation of atmospheric nitrogen in the target legume (Figueiredo *et al.* 2001). In this sense, it is of great importance to evaluate the potential of native strains, with the consequent selection of those most adapted to high temperature regions, such as those found in the Brazilian Northeast.

1.4. Nitrogen fertilization and seed quality

Seed is one of the main inputs in agriculture and its quality is one of the key factors in the establishment of any crop.

Seed quality can be expressed by the interaction of four components: genetic, physical, health and physiological (Ambrosano et al., 1999). According to Vieira et al. (1993), the physiological component can be influenced by the environment in which the seeds are formed. Therefore, germination and vigor must be taken into account, in an attempt to differentiate between seeds with greater physiological potential, depending on the cultural treatments applied, such as mineral fertilization (Andrade et al., 1999).

The availability of the nutrients provided by this fertilization influences the production and quality of the plants, affecting the formation of the embryo and reserve organs, as well as the chemical composition and, consequently, the metabolism and vigor of the seeds (Carvalho & Nakagawa, 2000). In addition, adequate fertilization can prevent some anomalies in seedling development, which are the most common manifestations of mineral deficiencies (Wiriglyr et al., 1984).

Nitrogen is a primary macronutrient, essential for plants because it participates in the formation of proteins, amino acids and other important compounds in plant metabolism. Its absence blocks the synthesis of cytokinin, the hormone responsible for plant growth, causing a reduction in plant size and consequently a reduction in the economic production of seeds (Mengel & Kirkby, 1987).

According to Carvalho & Nakagawa (2000), N can influence the physiological quality of seeds, but its effects vary according to environmental conditions and the stage of plant development at which the fertilizer is applied.

The effects of this nitrogen fertilization on the physiological quality of seeds are somewhat controversial in various crops. In the case of common beans, Carvalho et al. (1999) found an influence of nitrogen sources and application methods on the physiological quality of the seeds. However, Paulino et al. (1999) found no significant differences between nitrogen sources and ways of spreading nitrogen on the physiological quality of bean seeds.

Bassan et al. (2001), studying the cultivar Pérola "in winter", found increasing germination values of over 90% with the application of nitrogen up to the dose of 90 kg ha^{-1} of N in top dressing, in the absence of foliar fertilization with molybdenum. These authors also reported that the dose of 58 kg ha^{-1} of N allowed the maximum germination value for the accelerated ageing test (81%) in the treatment referring to the application of cover nitrogen fertilization without inoculation.

However, Ambrosano et al. (1999) and Carvalho et al. (2001) found no positive effect of nitrogen doses and application times on germination and vigor (accelerated aging) for the

IAC Carioca cultivar, "in winter". Crusciol et al. (2003), in a study carried out during the "spring" period with this cultivar, also found no significant effect of nitrogen doses, either in sowing or top dressing, on germination, which was above 90%.

1.5. Bibliographical references

AMBROSANO, E.J.; AMBROSANO, G.M.B.; WUTKE, E.B.; BULISANI, E.A.; MARTINS, A.L.M. & SILVEIRA, L.C.P. Effects of nitrogen and micronutrient fertilization on seed quality of the bean cultivar IAC - Carioca. **Bragantia**, Campinas, v.58, n.2, p.393-399, 1999.

ANDRADE, M.J.B. Climate and soil. In: VIEIRA, C.; PAULA JÙNIOR, T.J. de; BORÉM, A. (Ed.). **Beans:** general aspects and cultivation in the state of Minas Gerais. Viçosa: Editora da UFV, 1998. p.83-97.

ANDRADE, W.E.B.; SOUZA-FILHO, B.F.; FERNANDES, G.M.B.; SANTOS, J.G.C. Evaluation of the productivity and physiological quality of bean seeds submitted to NPK fertilization. In: TECHNICAL COMMUNICATION. Niteroi: PESAGRO- RIO, n.248, 5p., 1999.

ANDRIOLO, J.; PEREIRA, P. A. A.; HENSON, R. A. Variability among lines of wild forms for characteristics related to biological N2 fixation. **Pesquisa Agropecuària Brasileira**, Brasilia, v. 29, n. 6, p. 831-837, jun. 1994.

ARAÙJO, J.P.P. de; RIOS, G.P.; WATT, E.E.; NEVES, B.P. das; FAGERIA, N.K.; OLIVEIRA, I.P. de; GUIMARAES, C.M.; SILVEIRA FILHO, A. **Culture of caupi, *Vigna unguiculata (L.)* Walp.** Description and technical recommendations for cultivation. Goiânia: Embrapa-CNPAF, 1984. 82p. (Embrapa-CNPAF. Technical Circular, 18).

Araùjo, J.P.P.; Watt, E.E. O caupi no Brasil. 1.ed. Brasilia: EITA/EMBRAPA, 1988. 722p.

ARRESE-IGOR, C.; ESTAVILLO, J.M.; PENA, J.I., GONZALEZ-MURUA, C.; APARICIO-TEJO, P.M. Effect of low nitrate supply on nitrogen fixation in alfalfa root nodules induced by Rhizobium meliloti strains with varied nitrate reductase activity. **Journal of Plant Physiology**, Elsevier, v.135, p.207-211, 1989.

ASSAD, M.L.L.; ALMEIDA, J. Agriculture and sustainability: context, challenges and scenarios. **Ciência & Ambiente**, Santa Maria, RS, n. 29, p.15-30, 2004.

BASSAN, D.A.Z.; ARF, O.; BUZETTI, S.; CARVALHO, M.A.C.; SANTOS, N.C.B.; SA, M.E. Seed inoculation and application of nitrogen and molybdenum in winter bean crops: production and physiological quality of seeds. **Revista Brasileira de Sementes,** Brasilia, v.23, n.1, p.76-83, 2001.

BLOOM, A J.; SUKRAPANNA, S. S.; WARNER, R. L. Root respiration associated with ammonium and nitrate absorption and assimilation by barley. **Plant Physiology**, Lancaster, v. 99, p. 1294-1301. 1992.

BODDEY, L.H.; HUNGRIA, M. Phenotypic grouping of Brazilian *Bradyrhizobium* strains which nodulate soybean. **Biology and Fertility of Soils**, v.25, p.407-415, 1997.

BRANDÂO JUNIOR, O. & M. HUNGRIA. Effect of doses of peat inoculants on biological nitrogen fixation by the soybean crop. **Brazilian Journal of Soil Science**. Viçosa-MG, v.24, p. 527-535. 2000.

CARVALHO, M.A.C. et al. Influence of nitrogen sources and application methods on the physiological quality of winter bean seeds (*Phaseolus vulgaris* L.). **Informativo ABRATES**, Londrina, v.9, n.1/2, p.118, 1999.

CARVALHO, M.A.C.; ARF, O.; SA, M.E.; BUZETTI, S.; SANTOS, N.C.B.; BASSAN, D.A.Z. Productivity and quality of bean seeds (*Phaseolus vulgaris* L.) under the influence of nitrogen installments and sources. **Revista Brasileira de Ciência do Solo,** Viçosa, v. 25, n.3, p. 617-624, 2001.

CARVALHO, N.M.; NAKAGAWA, J. **Seeds: science, technology and production.** Jaboticabal: FUNEP, 2000. 588p.

CASTELLÓN, R.E.R; ARAÙJO, F.M.M.C.; RAMOS, M.V.; ANDRADE NETO, M.; FREIRE FILHO, F.R.; GRANGEIRO, T.B.; CAVADA, B.S. Elemental composition and characterization of the lipid fraction of six caupi cultivars. Revista Brasileira de Engenharia Agricola e Ambiental, Campina Grande, PB, v.7, n.1, p.149-153, 2003.

CASTRO, I.V. Ecotoxicological effects of heavy metals on biological nitrogen fixation in industrially contaminated soils. **Silva Lusitana**, Lisboa. Portugal, v. 8, p. 165-194. 2000.

CECCATTO, V. M.; GOMES, J. E.; SARRIÉS, G. A.; MOON, D. H.; TSAI, S. M. Effects of host plant origin on noduling activities and nitrogen fixation in *Phaseolus vulgaris* L. **Plant and Soil**, Dordrecht, v. 204, p. 79-87, 1998.

CENTRO INTERNACIONAL DE AGRICULTURA TROPICAL (Cali, Colombia). **Frijol production systems**. Cali, 1975. 64 p.

CHEN, W.X.; TAN, Z.Y.; GAO, J.L.; LI, Y.; WANG, E.T. *Rhizobium hainanense* sp. nov., isolated from tropical legumes. **International Journal of Systematic Bacteriology**, Iowa. v.47, p.870-873, 1997.

CHEN, W.X.; YAN, G.H.; LI, J.L. Numerical taxonomic study of fast-growing soybean rhizobia and a proposal that *Rhizobium fredii* be assigned to *Sinorhizobium* gen. nov.

International Journal of Systematic Bacteriology, Iowa, v.38, p.392-397, 1988.

CRUSCIOL, C.A.C.; LIMA, E.V.; ANDREOTTI, M.; NAKAGAWA, J.; LEMOS, L.B.; MARUBAYASHI, O.N. Effect of nitrogen on the physiological quality, productivity and characteristics of bean seeds. **Revista Brasileira de Sementes**, Brasilia, v. 25, n. 1, p. 108-115, 2003.

DUARTE, L.M.G. Globalizaçao, agricultura e meio ambiente: o paradoxo do desenvolvimento dos cerrados. In: Laura M. G. Duarte and Braga, M. L. de S. (eds.). **Tristes Cerrados**: society and biodiversity. Brasilia: Paralelo 15. 1998. p. 11-22.

FIGUEIREDO, M.V.B., EGiDIO, B.N.; BURITY, H.A.. Water stress response on the enzymatic activity in cowpea nodules. **Brazilian Journal of Microbiology**, Sao Paulo, v. 32, p. 195-200. 2001.

FRANCO, M. C.; CASSINI, S.T.A.; OLIVEIRA, V.R.; VIEIRA, C.; TSAI, S.M. Nodulation in bean cultivars from the Andean and Mesoamerican gene pools. **Pesquisa agropecuària brasileira**, Brasilia, v. 37, n. 8, p. 1145-1150, aug. 2002

FREIRE FILHO, F.R.; RIBEIRO, V. Q.; BARRETO, P. D.; SANTOS, C. A. **Genetic improvement of caupi (Vigna unguilculata (L) Walp.) in the Northeast region**. In: WORKSHOP, 1998. [S.1.]: Embrapa Semi-Arido, 1998.

JESUS, E.C.; SCHIAVO, J.A.; FARIA, S.M. Dependence on mycorrhizae for the nodulation of tropical leguminous trees. **Revista Arvore**, Viçosa-MG, v.29, n.4, p.545-552, 2005.

JORDAN, D.C. Transfer of *Rhizobium japonicum*, Bucchanan 1980 to *Bradyrhizobium* gen. nov., a genus of slow-growing, root nodule bacteria from leguminous plants. **International Journal of Systematic Bacteriology**, Iowa, v.32, p.136-139, 1982.

KIPE-NOLT, J. A.; MONTEALEGRE, M. C. M.; THOME, J. Restriction of nodulation by the broad host range of *Rhizobium tropici* strain CIAT 899 in wild accessions of *Phaseolus vulgaris* L. **New Phytologist**, New York, v. 120, n. 4, p. 489-494, 1992.

KOLLING, J.; SCHOLLES, D.; SELBACH, P.A. Selection of Rhizobium strains for subterranean clover, alfalfa and gherkin. **Agronomia Sulriograndense**, Porto Alegre, v.19, n.2, p.103-111, 1983.

KUYKENDALL, L.D.; SAXENA, B.; DEVINE, T.E.; UDELL, S.E. Genetic diversity in *Bradyrhizobium japonicum* Jordan 1982 and a proposal for *Bradyrhizobium elkanii* sp. nov. **Canadian Journal of Microbiology**, v.38, p.501-505, 1992.

LAJUDIE, P. de; WILLEMS, A.; POT, B.; DEWETTINCK, D.; MASTROJUAN, G.; NEYRA,

M.; COLLINS, M.D.; DREYFUS, B.; KERSTERS, K.; GILLIS, M. Polyphasic taxonomy of rhizobia: emendation of the genus *Sinorhizobium* and description of *Sinorhizobium meliloti* comb. nov, *Sinorhizobium saheli* sp. nov., and *Sinorhizobium teranga* sp. nov. **International Journal of Systematic Bacteriology**, Iowa, v.44, p.715-733, 1994.

MARTINS, L.M.V. **Ecological and physiological characteristics of caupi rhizobia (*Vigna unguiculata* (L.) Walp) isolated from soils in the Northeast region of Brazil**. 1996. 213p. Dissertation (Master's Degree) - Federal University of Rio de Janeiro, Seropédica.

MENGEL, K. and KIRKBY, E.A. **Principles of Plant Nutrition.** Bern: International Potash Institute. 1987. 687 p.

MIRANDA, P.; COSTA, A.F.; OLIVEIRA, L.R.; TAVARES, J.A.; PIMENTEL, M.L.; LINS, G.M.L. Behavior of cultivars of *Vigna unguiculata* (L) Walp. in single and intercropped systems. IV - erect and semi-erect types. **Pesquisa Agropecuària Pernambucana**, Recife, v. 9, n. especial, p. 95-105, 1996.

MOREIRA, F.M.S.; SIQUEIRA, J.O. **Microbiologia e bioquimica do solo**. Lavras: Federal University of Lavras, 2002. 625p.

MOTTA, VT. Bioquimica bàsica. 2. ed. Sao Paulo: Autolab, 2003. 365p.

NEVES, M.C.P.; RUMJANEK, N.G. Diversity and adaptability of soybean and cowpea rhizobia in tropical soils. **Soil Biology and Biochemistry**, v.29, p.889-895, 1997.

OLIVEIRA, A.P.; TAVARES SOBRINHO, J.; NASCIMENTO, J.T; ALVES, A.U; ALBUQUERQUE, I.C.; BRUNO, G.B. Evaluation of lines and cultivars of cowpea in Areia, PB. **Horticultura Brasileira**, Brasilia, v. 20, n. 2, p. 180-182, June 2002.

PAFFETTI, D.; DAGUIN, F.; FANCELLI, S.; GNOCCHI, S.; LIPPI, F.; SCOTTI, C.; BAZZICALUPO, M. Influence of plant genotype on the selection of nodulating Sinorhizobium meliloti strains by Medicago sativa. **Antonie van Leeuwenhoek**, v. 73, p. 3-8, 1998.

PAULINO, H.B. et al. Influence of the installment of two nitrogen sources, in coverage and via fertigation, on the physiological quality of bean seeds. **Informativo ABRATES**, Londrina, v.9, n.1/2, p.55, 1999.

RUMJANEK, N. G.; MARTINS, L. M. V.; XAVIER, G. R.; NEVES, M. C. P. Biological nitrogen fixation. In: FREIRE FILHO, F. R.; LIMA, J. A. de A.; RIBEIRO, V. Q. (Ed.). **Cowpea:** technological advances. Brasilia, DF. Embrapa Informaçao Tecnolócica, Teresina. Embrapa Meio-Norte, 2005. p. 281-355.

SCHLESINGER, W. H. **Biogeochemistry: An Analysis of Global Change**, 2 ed. Academic Press, San Diego, CA. 1997.

SCHUBERT, K.R. Products of biological nitrogen fixation in higher plants: synthesis, transport and metabolism. **Annual Review of Plant Physiology and Plant Molecular Biology**, Palo Alto, California, v. 37, p. 537-574, 1986.

SILVA, P.S.L.; OLIVEIRA, C.N. Yields of green and mature beans from caupi cultivars. **Horticultura Brasileira,** Brasilia, v. 11, n. 2, p. 133- 135, 1993.

SINGH, S. P.; GEPTS, P.; DEBOUCK, D. G. Races of common bean (*Phaseolus vulgaris*, Fabaceae). **Economic Botany**, New York, v. 45, p. 379-396, 1991.

SOARES, A. L. L.; PEREIRA, J.P.A.R.; FERREIRA, P.A.A.; MOREIRA, F. M. S.; ANDRADE, M. J. B. Nodulation and productivity of cowpea cv. BR 14 Mulato by selected rhizobium strains in Perdôes-MG. In: CONGRESSO NACIONAL DE FEIJÂO-CAUPI, 1., 2006, Teresina. **Proceedings**... Teresina: Embrapa, 2006a.

Available at < http://www.cpamn.embrapa.br/anaisconac2006> Accessed on January 26, 2007.

SOARES, A.L.; PEREIRA, J.P.A.R.; FERREIRA, P.A.A.F.; VALE, H.M.M.; LIMA, A.S.; ANDRADE, M.J.B.; MOREIRA, M.S.. Agronomic efficiency of selected rhizobia and diversity of native nodulating populations in Perdoes (MG). I - caupi. **Revista Brasileira de Ciência do Solo**, Viçosa-MG, v. 30, p.795-802, 2006.

TAIZ, L.; ZEIGER, E. **Plant physiology.** 3.ed. Porto Alegre: Artmed, 2004. 719p.

TEANEY III, G.B.; FUHRMANN, J.J. Soybean response to nodulation by bradyrhizobia differing in rhizobitoxine phenotype. **Plant and Soil**, Dordrecht, v.145, p.275-285, 1992.

TWARY, S.N.; HEICHEL, G.H. Carbon costs of dinitrogen fixation associated with dry matter accumulation in alfalfa. **Crop Science**, Madison, v.31, p.985-992, July/Aug. 1991.

VANCE, C.P.; HEICHEL, G.H.; PHILLIPS, D.A. Nodulation and symbiotic dinitrogen fixation. In: HANSON, A.A.; BARNES, O.K.; HILL JUNIOR, R.R. (Eds.). Alfalfa and alfalfa improvement. Madison: **American Society of Agronomy**, 1988. p.303-332.

VARGAS, M.A.T. & HUNGRIA, M. **Biologia dos solos dos cerrados**. Embrapa, Planaltina. 1997. 524 p.

VIEIRA, R.F.; VIEIRA, C.; CALDAS, M.T. Behavior of cowpeas in spring-summer in the Zona da Mata of Minas Gerais. **Pesquisa Agropecuâria Brasileira**, Brasilia, v.35, n.7, p.1359-1365, 2000.

VIEIRA, R.F.; VIEIRA, C.; RAMOS, J.A.O. **Bean seed production.** Viçosa: EPAMIG/EMBRAPA, 1993. 131p.

WIRIGLYR, C.W.; DU CROS, D.L.; MOSS, H.J. et al. Effect of sulfer deficiency on wheat evaluation. In: **SULFUR in Agriculture**. Washington: The Sulfur Institute, 1984. p.2-7.

XAVIER, G.R; MARTINS, L.M.V.; RIBEIRO, J.R.A.; RUMJANEK, N.G. Symbiotic specificity between rhizobia and cowpea accessions of different nationalities. **Caatinga,** Mossoró, v.19, n.1, p.25-33, January/March 2006.

ZILLI, J.E. **Characterization and selection of rhizobium strains for inoculation of caupi (*Vigna unguiculata* (L.) Walp.) in Cerrado areas**. 2001. 112p. Dissertation (Master's Degree) - Federal Rural University of Rio de Janeiro, Seropédica.

ZILLI, J.E.; VALICHESKIR.R.; RUMJANEK, N.G.; SIMOES-ARAÙJO, J.L.; FREIRE FILHO, F.R. E NEVES, M.C.P. Symbiotic efficiency of *Bradyrhizobium* strains isolated from Cerrado soil in caupi. **Pesquisa Agropecuària Brasileira**, Brasilia, v.41, n.5, p.811-818, May 2006.

CHAPTER II. EFFICIENCY OF RHIZOBIUM STRAINS IN THE PRODUCTION AND QUALITY OF COWPEA SEEDS - 2005 CROP YEAR

2.1 Introduction

The cowpea [*Vigna unguiculata* (L.) Walp.], also known as the macassar bean or string bean, is of great socio-economic importance for the North and Northeast regions of Brazil, especially for the population with lower purchasing power, making it one of the most important components of their diet, as its seeds are one of the main sources of protein and energy for humans. In the 1990s, the crop produced an average of 430,000 tons per year^{-1} , harvested from around 1,300,000 hectares. It is estimated that caupi is responsible for creating 1.36 million jobs every year and for feeding more than 23 million Brazilians, with a production value of around 250 million dollars (Freire Filho et al., 2005).

Despite its socio-economic importance, historically this crop has had low productivity, according to Alcântara et al. (2006), around 300 to 400 kg ha^{-1} , largely due to the growing conditions without the adoption of advanced technologies. However, this productivity could be increased through the use of efficient rhizobial inoculants, which could meet the plant's nitrogen needs (Zilli, 2001), given that nitrogen (N) is one of the most limiting nutrients for cereal production in tropical climate regions, and its biological fixation is of great importance for plant production.

The symbiotic association between legumes and bacteria of the genus *Rhizobium* culminates in the formation of specialized organs (nodules) in which the bacteria are able to convert atmospheric N_2 into ammonia, the form used by the plants.

This process can be an alternative for the total or partial replacement of nitrogen fertilizers, as long as it supplies the crop with the nitrogen necessary for its growth and development, as well as reducing production costs and saving fossil fuels used to manufacture nitrogen fertilizers (Soares et al., 2006).

One of the most successful examples of biological nitrogen fixation (BNF) is the case of soybeans in Brazil, where the use of inoculants with *Bradyrhizobium* has led to annual savings of US$ 3 billion in nitrogen fertilizers (Hungria et al., 2005). For cowpeas, estimates of the contribution of BNF are in the order of US$ 13 million for the Northeast region of Brazil alone (Rumjanek et al., 2005), since, according to Pereira et al, (2006), most of the inoculants produced and imported in Brazil are intended for soybean crops (around 99%), with the remainder (1%) going to other legume species, including cowpeas, which indicates the need for greater dissemination of this low-cost biotechnology.

The selection of new strains capable of fixing greater amounts of N2 when in symbiosis with caupi is extremely important for the production of inoculants. However, the strains selected in the laboratory and greenhouse may not reach their maximum fixing potential in the field, due, among other factors, to competition with the native population established in the soil and adaptation to local environmental conditions (Soares et al., 2006).

The aim of this study was to evaluate the efficiency of new rhizobium strains on the production and physiological quality of cowpea seeds.

2.2 Material and Methods

The experiment was carried out in partnership with Embrapa Agrobiologia and conducted in the experimental area of Campus II of the State University of Paraiba (UEPB), located in the municipality of Lagoa Seca - PB, from August to November 2005. The evaluation of the inoculants followed the recommendations of RELARE (Network of Laboratories for the Recommendation, Standardization and Dissemination of Technology for Microbial Inoculants of Agricultural Interest). The monthly rainfall, according to data obtained from the meteorology department at this campus, is shown in Figure 1.1.

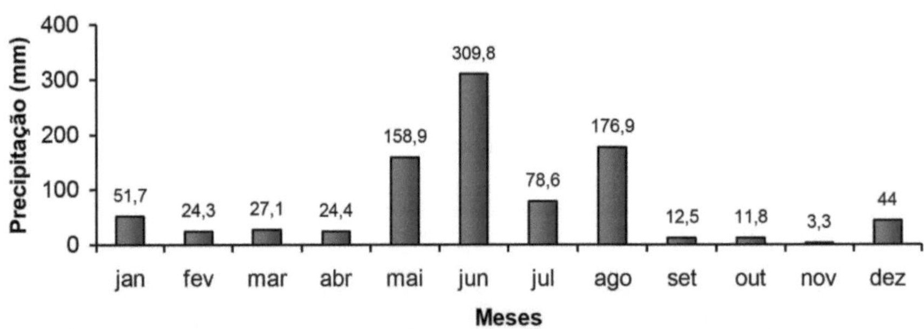

Figure 1.1 Monthly rainfall at the UEPB Campus II experimental unit (2005).

The soil was analyzed at the Department of Soils and Plant Nutrition at Esalq/University of São Paulo. It was classified as Neossolo Regolitico, and the results of the chemical analysis for fertility purposes on samples taken between 0.0 and 0.2 m are shown in Table 1.1. According to the limits for interpreting chemical characteristics (Raij et al., 1997), some chemical elements showed medium levels (MO, P, B, Cu, Fe, Mn, Zn, K), while others showed low levels (Ca, Mg, Al, SB, V).

Table 1.1 Chemical properties of the soil at the experimental unit of the Colégio Agricola Assis Chateaubriand (2005).

pH	MO	P	B	Cu	Fe	Mn	Zn
($CaCl_2$)	(g/dm)^3		-		(mg/dm)^3 ---------	-	
4,5	17,0	20,0	0,3	0,5	62,0	5,1	0,9

K	Ca	Mg	Al	H+Al	SB	T	V
			-(mmolc/ dm^3)-		-		(%)
1,8	6,0	2,0	2,0	22,0	9,8	31,8	31,0

SB- sum of bases; T- cation exchange capacity. Department of Soils and Plant Nutrition, Esalq. University of Sao Paulo.

Soil preparation consisted of manual weeding to eliminate weeds, followed by harrowing to make it easier to open the holes, fertilizing and leveling the soil. After this phase, the soil was limed to correct the pH and all the plots were given a foundation fertilizer, which consisted of 80 kg ha^{-1} of P_2O_5 (simple superphosphate), 30 kg ha^{-1} of K_2O (potassium chloride) and 10 kg ha^{-1} of micronutrients in the form of H_3BO_3, $CuSO_4$, $ZnSO_4$ and $MnSO_4$, respectively, applied to the previously opened holes.

The experimental plots were 4 m x 6 m, with a total of twelve 4 m rows each. The holes were dug manually at a spacing of 0.5m x 0.5m. In order to collect data for statistical analysis, a useful area was determined in each plot, which consisted of six central rows, with 1 m on each side, which established a space of 6 m^2.

When sowing, five seeds were distributed per hole, from lots obtained by farmers in the region, following RELARE's recommendations for this project. The cowpea seeds, cultivar Corujinha, were inoculated with various rhizobium strains, namely: BR 3301 (INPA 03-11B or SEMIA 6463) and BR 3302 (UFLA 3-84 or SEMIA 6461), from the culture collection of the Soil Microbiology Laboratory at the Federal University of Lavras; BR 3267 (SEMIA 6462), BR 3299 and BR 3262, from the culture collection of Embrapa Agrobiologia. It should be noted that, at the time of experimentation, two of the strains mentioned (BR 3299 and BR 3262) were not yet officially recommended for cowpea inoculation and, in this work, their development was evaluated and compared with that of the other strains currently on the inoculant market for this crop. Recently, the Ministry of Agriculture, Livestock and Supply (MAPA) authorized the use of strain BR 3262 as a cowpea inoculant (Brasil, 2011).

To inoculate these seeds, 30 grams of rhizobium mixed in a solution containing 20 ml of water and 30 grams of sugar were used for every 600 grams of seeds. To ensure that the inoculant adhered better to the seeds, they were mixed manually and left to dry in the shade for 30 minutes before sowing.

With regard to fertilization, the following doses of nitrogen fertilizer (urea) were used for comparison with the inoculation of the strains: 50 kg ha^{-1} of nitrogen (at planting) and 80 kg ha^{-1} of nitrogen (40 kg ha^{-1} at planting and 40 kg ha^{-1} at 35 days after sowing).

One week after germination, thinning was carried out, leaving 3 (three) plants per hole and a total of 72 (seventy-two) plants in the useful area of each plot. Weeding was also carried out whenever necessary in order to keep the area clean throughout the experimental period.

To evaluate the treatments used, the following variables were analyzed:

a) Number of nodules - determined by manually counting the nodules of ten plants, collected with intact roots, from the third row on each side of the plot, 35 days after the seedlings emerged;

b) Dry weight of nodules - based on the weight of the nodules removed to count the nodulation, which were dried in a forced air circulation oven at a temperature of 65° C for 48 h and the result expressed in mg plant ;$^{-1}$

c) Nodule weight - this was assessed using the ratio between the dry weight of the nodules and the number of nodules, with the result expressed in mg nodule ;$^{-1}$

d) Relative efficiency of the plot - calculated by dividing the dry weight of the aerial part of the 80 kg ha^{-1} N treatment by the dry weight of the aerial part of the other treatments, and multiplying by 100;

e) Dry weight of the aerial part and root - the aerial part (stem and leaves) and root of three plants collected from the second row of the plot at 35, 55 and 75 DAE were dried in a forced-air oven at 65° C for 72 hours;

f) Number of pods per plant - derived from the ratio of the total number of pods in the useful area and the number of plants in this area;

g) Number of seeds per pod - based on the average of ten pods taken from the useful area;

h) Pod length - the ten pods used to determine the previous variable were measured using a ruler graduated in centimeters;

i) Seed yield - obtained from the production of the useful area, corrected to 13% humidity and expressed in kg ha ;$^{-1}$

j) Weight of 1,000 seeds - determined using the average of eight replicates of 100 seeds per treatment (Brasil, 1992);

l) Seed yield per plant - established by the ratio between the yield of the useful area, corrected to 13% moisture, and the number of plants in it;

m) Number of seeds per plant - calculated by dividing the number of seeds in the useful area by the number of plants;

n) Seed length - the longest part of the seed was determined using a caliper;

o) Seed width - the length from the hilum to the opposite side of the seed was measured using a caliper;

p) Seed thickness - determined by the lengths transverse to the width of the seed;

q) Germination - seed germination was assessed in accordance with the Rules for Seed Analysis (Brazil, 1992), using the standard germination test;

r) Dry weight of seedlings - the seedlings from the germination test were placed in paper bags and dried in an oven at 65° C until they reached a constant weight, based on Nakagawa (1999);

s) Emergence of seedlings in sand - plastic trays containing sterilized sand and moistened with water equivalent to 60% of the holding capacity were distributed in the greenhouse, where four replications of 50 seeds were sown to carry out this test (Brasil, 1992);

t) First seedling emergence count - was obtained from the emergence data of the first seedlings to emerge from the aforementioned test;

u) speed of emergence index in sand - was also carried out using the seedling emergence test in sand, where daily counts of emerged seedlings were carried out from the first count until the number became constant. The emergence speed index (ESI) was calculated using the formula proposed by Maguire (1962).

The experimental design used was a randomized block design with four replications. The treatments were represented by rhizobium strains (BR 3301, BR 3302, BR 3267, BR 3299 and BR 3262), nitrogen fertilization (50 kg ha^{-1} , 80 kg ha^{-1}) and a control treatment (no inoculation and no mineral N). The statistical analysis program SISVAR, version 4.6, was used to analyze the data. The means of the treatments were compared using the Scott-Knott test at 5% probability.

2.3 Results and Discussion

The summary of the analysis of variance can be found in the Appendix (A 1), while the averages of the variables relating to the symbiotic efficiency of the strains in nodulation are shown in Table 1.2.

Table 1.2 Average values of nodule number (NN), nodule dry weight (PSN), nodule weight (PN) and relative plot efficiency (EfR) from cowpea plants subjected or not to inoculation and nitrogen fertilization.

Treatments	NN (ud/plant)	PSN (mg/plant)	PN (mg/nod.)	EfR (%)
BR 3301	21,50 a	75,10 a	3,50 b	131,00 a
BR 3302	15,75 a	79,02 a	4,75 a	121,00 a
BR 3262	17,75 a	84,32 a	5,00 a	117,50 a
BR 3267	20,50 a	88,47 a	4,50 a	122,25 a
BR 3299	14,75 a	54,40 b	3,75 b	129,75 a
80 kg ha^{-1}	15,75 a	63,45 a	4,25 a	100,00 a
50 kg ha^{-1}	11,50 a	38,95 b	3,50 b	96,00 a
Control	12,00 a	22,37 b	2,00 c	93,00 a
CV (%)	35,70	40,41	26,15	29,64

Averages followed by the same letter do not differ according to the Scott-Knott test at 5%.

The number of nodules (Table 1.2) found on the roots of plants from seeds inoculated with strains BR 3299 and BR 3262 were statistically similar to those produced by the strains officially recommended (Diário Oficial da Uniâo, 2006) for this crop (BR 3301, BR 3302 and BR 3267), which in turn did not differ from the other treatments.

The dry weight of the nodules showed significant differences when the strains were inoculated. With the exception of BR 3299, the other inoculated strains resulted in plants with a higher dry weight when compared with the application of 50 kg ha^{-1} of nitrogen and with native rhizobia, represented here by the control treatment (Table 1.2).

When the seeds were inoculated with strains BR 3302, BR 3262 and BR 3267, they gave rise to plants with the highest average nodule weights, but they did not differ from those that received mineral fertilization (80 kg ha^).[1]

On the other hand, the relative efficiency of the plots was not affected by the treatments at a 5% significance level.

The various environmental factors that can negatively affect the nodulation capacity and nitrogen fixation of rhizobium populations in soils include pH, humidity, temperature, biological factors, salinity, nitrates, pesticides, herbicides and heavy metals (Castro, 2000). In addition to these factors, the introduction of efficient inoculants can also be hampered by native strains which are generally very competitive (Santos et al., 2005).

Thus, the number of nodules found on the plants grown under the control treatment (without inoculant and nitrogen fertilizer) indicates the presence of native strains in the soil, which, although capable of supplying the plants with the nitrogen fixed by the rhizobium-caupi symbiosis, may have limited the establishment of the inoculated strains, as observed by Pereira et al. (2008).

As found in this study, the presence of nodulation in cowpea plants that did not receive inoculation treatments with rhizobium strains was also observed by other authors (Xavier et al., 2006; Gualter et al., 2008).

Lima et al. (2005) analyzed the phenotypic diversity and symbiotic efficiency of *Bradyrizhobium* spp. strains in Amazonian soils and found significant differences in the relative efficiency of plots cultivated with cowpea. Among the most efficient strains found by these authors, BR 3301 and BR 3302 also stood out.

Other promising results were also verified by Melo and Zilli (2009) when they used strains BR 3267 and BR 3262, as they found that the number of nodules and the mass of dry nodules found on cowpea plants were higher than those found on plants in the absolute control treatment and on plants fertilized with urea (50 kg ha^{-1} of N).

Evaluating the number and dry weight of nodules are some of the criteria used by RELARE to assess symbiotic efficiency between rhizobia and legumes. In this case (Table 1.2), specifically, the performance of strain BR 3262 was similar to the results obtained by the strains recommended for this crop and to results from other studies (Lacerda et al., 2004 and Soares et al., 2006).

In experiments with the cowpea cultivar BRS Amapà, Zilli et al. (2006) used the same treatments as this work and found that in the 2005 harvest, the number of nodules and their dry weight were not influenced by these treatments when the crop was grown in a forest area in the state of Roraima. On the other hand, these variables were significantly affected in the same season when the cowpea plants were grown in the cerrado area of this state.

State. In the latter case, strain BR 3262 provided the best results in both variables analyzed.

In view of the above, there is a clear need to select strains adapted to each agro-ecological situation, since each of these strains responds differently to the environment in which it is introduced.

According to the average values for the dry weight of the aerial part and root (Table 1.3), no significant differences were found between the treatments used. This may be due to the good nutrient content found in the soil evaluated (Table 1), since according to Freire Filho et al. (2005), levels equal to or greater than 10 and 20 mg dm^{-3} for P and K, respectively, do not provide responses from cowpea plants to nutrient application.

The results of this work corroborate Morgado et al. (2006) who also found no influence of rhizobium inoculation on the dry weight of cowpea plants, when the seeds that generated these plants were inoculated with strains BR 3301, BR 3302, BR 3267 and BR 3299. On the other hand, work carried out by Gualter et al. (2006), in Terezina - PI, showed that the dry weight of the aerial part and the roots were modified by the different strains (BR 3262, BR 3301, BR 3267, BR 3280, BR 3287 and BR 3299), with BR 3262 promoting the best results in both variables analyzed.

Table 1.3 Average values for aerial part dry weight (PSPA at 35, 55 and 75 DAE) and root dry weight (PSR at 35, 55 and 75 DAE) from cowpea plants subjected or not to inoculation and nitrogen fertilization.

	PSPA (g/plant)			**PSR** (g/plant)		
Treatments	35 D.A.E	55 D.A.E	75 D.A.E	35 D.A.E	55 D.A.E	75 D.A.E
BR 3301	1,25 a	3,35 a	5,69 a	0,30 a	0,94 a	0,61 a
BR 3302	1,19 a	3,20 a	9,60 a	0,31 a	0,96 a	0,71 a
BR 3262	1,10 a	2,28 a	8,42 a	0,36 a	0,87 a	0,75 a
BR 3267	1,38 a	2,73 a	7,37 a	0,40 a	0,85 a	0,83 a
BR 3299	1,09 a	2,68 a	6,13 a	0,34 a	0,96 a	0,70 a
80 kg ha^{-1}	1,34 a	3,39 a	7,97 a	0,32 a	0,88 a	0,80 a
50 kg ha^{-1}	1,06 a	2,30 a	9,52 a	0,33 a	0,64 a	0,90 a
Control	0,86 a	2,67 a	5,36 a	0,26 a	0,69 a	0,60 a
CV (%)	41,91	32,52	47,74	26,59	27,84	26,28

Averages followed by the same letter do not differ according to the Scott-Knott test at 5%.

The results found by the authors mentioned above also corroborate the research carried out by Zilli et al. (2006), when they found that strain BR 3262 promoted the greatest accumulation of dry weight of the aerial part of cowpea plants (35 D.A.E.) grown in cerrado soil.

In this sense, the selection of these strains is a stage of fundamental importance in the production of efficient commercial inoculants, seeking the best plant development in the field (Jesus et al., 2005).

Table 1.4 shows the average values for seed dimensions (length, width and thickness) and the weight of 1,000 seeds. No difference was observed between the treatments in terms of seed length. On the other hand, the width, thickness and weight of the seeds were affected by the treatments given to the crop.

Table 1.4 Average values for seed length (LS), seed width (LS), seed thickness (ES) and weight of 1,000 seeds (PS) from cowpea plants subjected or not to inoculation and nitrogen fertilization.

Treatments	CS (mm)	LS (mm)	ES (mm)	PS (g)
BR 3301	8,80 a	7,02 b	5,05 b	24,26 b
BR 3302	8,97 a	7,27 a	5,25 b	24,55 b
BR 3262	9,10 a	7,52 a	5,50 a	23,58 c
BR 3267	8,95 a	7,20 b	5,32 a	23,65 c
BR 3299	9,15 a	7,52 a	5,55 a	24,55 b
80 kg ha^{-1}	8,97 a	7,42 a	5,47 a	25,31 a
50 kg ha^{-1}	8,95 a	6,97 b	5,02 b	23,71 c
Control	9,20 a	7,40 a	5,42 a	23,74 c
CV (%)	5,29	3,12	4,61	2,19

Averages followed by the same letter do not differ according to the Scott-Knott test at 5%.

In the width evaluation, the cowpea plants from seeds inoculated with strains BR 3262, BR 3299 and BR 3302 or from nitrogen fertilization (80 kg ha^{-1}) and the control treatment achieved the best results when compared to the other treatments. On the other hand, strains BR 3262, BR 3299 and BR 3267 or the treatments fertilized with nitrogen (80 kg ha^{-1}) and the control treatment, produced seed-bearing plants with greater thickness.

As for the weight of 1,000 seeds, the treatment that received 80 kg ha^{-1} of nitrogen produced seeds with the highest weight, differing from the other treatments.

Larger seeds generally have better physiological quality, which can be advantageous under conditions of water stress or shading (White and Gonzâlez, 1990).

Although evidence indicates that small soybean seeds reduce emergence and give rise to shorter plants, the superiority of large seeds in terms of grain yield has not been sufficiently proven (Lima and Carmona, 1999)

In the bean crop (*Phaseolus vulgaris* L.), Perin et al. (2002) observed that plants from small seeds can compensate for lower initial growth at later stages of the cycle, ensuring similar yields to plants from large seeds.

Table 1.5 shows the parameters assessed after harvest, which shows that there was no significant effect of the treatments on these characteristics.

Table 1.5 Average values for seed yield (SR), seed yield per plant (RSP), number of seeds per plant (NSP), number of pods per plant (NVP), pod length (CV) and number of seeds per pod (NSV) of cowpea plants subjected or not to inoculation and nitrogen fertilization.

Treatments	**RS** (kg/ha)	**RSP** (g/plant)	**NSP** (ud.)	**NVP** (ud.)	**CV** (mm)	**NSV** (ud.)
BR 3301	364,33 a	4,96 a	20,00 a	1,62 a	186,2 a	13,00 a
BR 3302	303,37 a	3,03 a	12,50 a	1,37 a	188,1 a	11,75 a
BR 3262	409,74 a	4,43 a	18,50 a	1,97 a	184,6 a	12,50 a
BR 3267	360,10 a	4,35 a	18,50 a	2,00 a	179,8 a	12,50 a
BR 3299	376,65 a	4,32 a	17,50 a	2,00 a	184,2 a	11,75 a
80 kg ha^{-1}	344,03 a	5,49 a	21,75 a	1,75 a	187,9 a	12,50 a
50 kg ha^{-1}	378,57 a	6,55 a	27,50 a	3,17 a	185,4 a	11,75 a
Control	377,52 a	3,93 a	16,50 a	1,72 a	174,6 a	11,25 a
CV (%)	35,94	49,79	49,70	49,07	4,46	9,90

Averages followed by the same letter do not differ according to the Scott-Knott test at 5%.

It's worth noting that these results could have been a little more promising given that rainfall was already very scarce in the planting region close to the time when the pods formed (October) (Figure 1.1). However, the average yields of the treatments were within the average for the Northeast region, which is around 300 to 400 kg ha^{-1} (Alcântara, 2006).

Mendes (2000) presented results of agronomic efficiency in bean (*Phaseolus vulgaris* L.) showing that the response of this crop to inoculation, under field conditions, varies depending on the soil, the history of cultivation of the area and other biotic and abiotic factors.

In the semi-arid conditions of the state of Paraiba (Silva et al., 2006), the effect of four strains (BR 3267, INPA 3-11B or BR 3301, UFLA 3-84 or BR 3302 and UFRPE or NFB 700) inoculated into cowpea cultivar CNCx409- 11F was evaluated, where there were also no significant differences between these inoculants and the control treatments (with and without mineral nitrogen), but the seed yield (1,340 to 1,768 kg ha^{-1}) was much higher than that found in this work.

This difference in productivity between the yields of the authors mentioned above and those found in this experiment may have been influenced both by the low rainfall during the experimental phase (Figure 1.1) and also due to the productive capacity of the different cultivars used in the two studies, since differences in cowpea productivity ranging from 117.00 to 1,587.00 kg ha^{-1} have already been observed, depending on the genotype grown (Vieira et al., 2000).

However, other studies have already shown the positive effect of using rhizobium strains on the productivity of this crop (Lacerda et al., 2004; Lima et al., 2005; Pereira et al., 2006; Soares et al., 2006; Zilli et al., 2006).

Table 1.6 shows the results of the parameters relating to the quality of cowpea seeds when produced under the different treatments.

In terms of germination, the highest percentages were obtained in seeds from plants inoculated with strains BR 3302, BR 3262, BR 3301 and the control treatment (native strains). On the other hand, the seeds that gave rise to seedlings with the highest dry weight came from the treatment where nitrogen fertilizer was used (80 kg ha).$^{-1}$

Table 1.6 Average values of seed germination (GER), seedling dry weight (MSW), first count of seedling emergence in sand (PCEA), seedling emergence in sand (EPA) and speed of emergence in sand (IVE) of cowpea plants submitted or not to inoculation and nitrogen fertilization.

Treatments	**GER** (%)	**MSP** (mg/plant)	**PCEA** (%)	**EPA** (%)	**IVE**
BR 3301	98,0 a	55,50 b	89,0 a	97,00 a	19,12 a
BR 3302	100,0 a	57,50 b	92,0 a	95,00 a	18,90 a
BR 3262	99,0 a	51,48 b	90,0 a	93,00 a	18,50 a
BR 3267	95,5 b	54,30 b	88,0 a	92,00 a	18,25 a
BR 3299	94,5 b	52,96 b	95,0 a	95,00 a	19,00 a
80 kg ha^{-1}	95,5 b	72,71 a	86,0 a	93,00 a	18,37 a
50 kg ha^{-1}	93,0 b	50,97 b	90,0 a	92,00 a	18,32 a
Control	97,5 a	52,91 b	93,0 a	97,00 a	19,27 a
CV (%)	2,80	15,05	7,29	3,94	4,03

Averages followed by the same letter do not differ according to the Scott-Knott test at 5%.

When the data on the weight of 1,000 seeds is analyzed (Table 1.4), it can be seen that they weighed more when they were produced under mineral nitrogen fertilization at a dosage of 80 kg ha^{-1} . This suggests that seeds produced under suitable nutrient conditions have a higher density and consequently give rise to more vigorous seedlings.

These results corroborate Carvalho & Nakagawa (2000) when they postulated that the largest seeds or those with the highest density are those which normally have well-formed embryos and greater amounts of reserves, possibly being more vigorous.

In this regard, Gaspar and Nakagawa (2002) found that the germination and vigor of millet seeds (*Pennisetum americanum* L.) were influenced by the size of the seeds, since those of larger size showed better quality.

However, some studies on certain species have shown that the weight and size of the seeds did not influence the tests carried out in the laboratory and the development of seedlings in the field (Andrade et al., 1997; Martins et al., 1997; Martinelli-Seneme et al., 2001).

With regard to the first count, the percentage and speed of emergence of seedlings in sand, it can be seen that these were not influenced by the treatments used (Table 1.6).

Research into the influence of seed size and shape has produced divergent results with regard to the percentage and speed of emergence. Some authors also found no significant differences in the percentage of seedling emergence in the field when comparing flat and round maize seeds (Andrade et al., 1997) and between large and medium flat seeds (Silva & Marcos Filho, 1982).

2.4 Conclusions

The strains not yet officially recommended (BR 3262 and BR 3299) showed similar development to those already officially recommended for cowpea cultivation and mineral nitrogen fertilization at dosages of 50 and 80 kg ha ;$^{-1}$

Yield was not influenced by the treatments;

When strains BR 3302, BR 3262, BR 3301 and the control were used to grow cowpeas, seeds with a higher germination percentage were produced;

The cowpea seedlings were more vigorous (seedling dry weight) when they came from seeds of greater weight;

The results of the control treatment indicate the presence of an established population of native strains, with similar efficiency to the others observed in this study.

2.5 References

ALCANTARA, R. M. C. M.; FORTALEZA, J. M.; XAVIER, G. R.; SOUZA, J. S. Inoculation of cowpea [Vigna Unguiculata (L.) Walp.] with rhizobium BR 3267 in Teresina, Pi. In: CONGRESSO NACIONAL DE FEIJÂO-CAUPI, 1., 2006, Teresina.

Anais... Teresina: Embrapa, 2006. Available at < http://www.cpamn.embrapa.br/anaisconac2006> Accessed on January 26, 2007.

ANDRADE, R. V.; ANDREOLI, C.; BORBA, C. S.; AZEVEDO, J. T.; MARTINSNETTO, D. A.; OLIVEIRA, A. C. Efeito da forma e do tamanho da semente no desempenho no campo de dois genótipos de milho. **Revista Brasileira de Sementes,** Brasilia, v.19, n.1, p.62-65, 1997.

BRAZIL. Ministry of Agriculture and Agrarian Reform. **Rules for seed analysis**. Brasilia: SNDA/DNDV/CLAV, 1992. 365p.

BRAZIL. Ministry of Agriculture, Livestock and Supply. **Normative Instruction No. 13, of March 24, 2011**. Approves the rules on specifications, guarantees, registration, packaging and labelling of inoculants intended for agriculture, as well as the lists of microorganisms authorized and recommended for the production of inoculants in Brazil. Official Gazette of the Federative Republic of Brazil, 25. 2011. Section 1, p.3-7. Available at: <http://www.normasbrasil.com.br/norma/instrucao-normativa-13-201178540.html> Accessed on: August 2, 2015.

CASTRO, I.V. Ecotoxicological effects of heavy metals on biological nitrogen fixation in industrially contaminated soils. **Silva Lusit.**, 8: 165-194, 2000.

CARVALHO, N. M.; NAKAGAWA, J. **Seeds:** science, technology and production. 4 ed. Jaboticabal: FUNEP, 2000. 588p.

CARVALHO, N.M.; NAKAGAWA, J. **Seeds**: science, technology and production. 4.ed. Jaboticabal: FUNEP, 2000. 588p.

OFFICIAL GAZETTE OF THE UNION. Instruçâo normativa no 10, de 21 de março do 2006, seçâo 1. Available at < http://www.diario-oficial.com.br/servicos/diario_eletronico/index.asp> Accessed on January 30, 2007.

FREIRE FILHO, F.R.; LIMA, J.A.A.; RIBEIRO, V.Q. **Cowpea**: technological advances. Brasilia, DF: Embrapa Informaçâo Tecnològica, 2005. 519p.

GASPAR, C. M.; NAKAGAWA, J. Influence of size on the germination and vigor of millet seeds (*Penninsetum americum* (L.) Leeke). **Revista Brasileira de Sementes**, Brasilia, v. 24. n. 1, p.339-344, 2002.

GUALTER, R.M.R.; HENRIQUES NETO, D.; LEITE, L.F.C.; DANTAS, J.S.; HOLANDA NETO, M.R. Nodulation in cowpea inoculated with Bradyrhizobium spp. strains in a fluvial neosol in Teresina, PI. In: CONGRESSO NACIONAL DE FEIJÂO-CAUPI, 1., 2006,

Teresina. **Proceedings**... Teresina: Embrapa, 2006. Available at < http://www.cpamn.embrapa.br/anaisconac2006> Accessed on January 26, 2007.

GUALTER, R.M.R.; LEITE, L.F.C.; ARAÙJO, A.S.F.; ALCANTARA, R.M.C.M.; COSTA, D.B. Inoculation and mineral fertilization in cowpea: effects on nodulation, growth and productivity. **Sci. Agraria**, 9: 469-474, 2008.

JESUS, E.C.; SCHIAVO, J.A.; FARIA, S.M. Dependence on mycorrhizae for the nodulation of tropical leguminous trees. **Revista Arvore**, 29: 545-552, 2005.

HUNGRIA, M.; FRANCHINI, J.C.; CAMPO, R.J.; GRAHAM, P.H. The importance of nitrogen fixation to soybean cropping in South America. In: WERNER, D.; NEWTON, W.E. (Ed.). **Nitrogen fixation in agriculture, forestry, ecology, and the environment**. Dordrecht: Springer, p.25-42, 2005.

LACERDA, A.M.; MOREIRA, F.M.S.; ANDRADE, M.J.B. & SOARES, A.L.L. Effect of rhizobium strains on nodulation and productivity of cowpea. **Revista Ceres**, v. 51, p.67- 82, 2004.

LIMA, A. M. M. P.; CARMONA, R. Influence of seed size on soybean production performance. **Revista Brasileira de Sementes**, Brasilia, v. 21, n. 1, p. 157- 163, 1999.

LIMA, A.S.; PEREIRA, J.P.A.R.; MOREIRA, F.M.S. Phenotypic diversity and symbiotic efficiency of *Bradyrhizobium* spp. strains from Amazonian soils. **Pesquisa Agropecuâria Brasileira**, Brasilia, v.40, n.11, p.1095-1104, 2005.

MAGUIRE, J.D. Speed of germination-aid in selection and evaluation for seedling emergence and vigor. **Crop Science**, Madison, v.2, n.1, p.176-177,1962.

MARTINELLI-SENEME, A.; ZANOTO, M. D.; NAKAGAWA, J. Effect of seed shape and size on yield of maize cultivar AL-34. **Revista Brasileira de Sementes**, Brasilia, v. 23. n. 1, p.40-47, 2001.

MARTINS, C.O.A.; PADILHA, L.; FERREIRA, A. C. B.; MANTOVANI ALVARENGA, M.; DIAS, D. C. F. S. Influence of classification by treatment on germination and vigor of soybean seeds (*Glicine max* (L.) Merril) In: CONGRESSO BRASILEIRO DE SEMENES, 10, Foz do Iguaçu, Aug. 1997. **Informativo ABRATES**, Curitiba, v.27, n.1/2, p.169,1997.

MENDES, I. C. IX RELARE - **Circular No. 3** - Feb. 2000. Available at <http://www.hps.Org/hpsnews/19065.html>. Accessed on 25 Jan. 2007.

MORGADO, L.B.; MARTINS, L.M.V.; XAVIER, G.R. and RUMJANEK, N.G. Evaluation of the potential of rhizobium strains to fix nitrogen associated with cowpea in Petrolina-PE. In:

CONGRESSO NACIONAL DE FEIJÂO-CAUPI, 1., 2006, Teresina. **Anais**... Teresina: Embrapa, 2006. Available< http://www.cpamn.embrapa.br/anaisconac2006> Accessed on January 26, 2007.

NAKAGAWA, J. Vigor tests based on seedling performance. In: KRZYZANOWSKI, F.C., VIEIRA, R.D., FRANÇA NETO, J.B. **Seed vigor**: concept and tests. Londrina: ABRATES. 1999. 218p.

PEREIRA, J. P. A. R.; FERREIRA, P. A. A.; VALE, H. M. M.; NOGUEIRA, C. O. G.; SOARES, A. L. L.; MOREIRA, F. M. S.; ANDRADE, M. J. B. Nodulation and productivity of cowpea cv. Poços de Caldas by selected rhizobium strains in Iguatama-MG. In: CONGRESSO NACIONAL DE FEIJÂO-CAUPI, 1., 2006, Teresina. **Proceedings**... Teresina: Embrapa, 2006. Available at < http://www.cpamn.embrapa.br/anaisconac2006> Accessed on January 26, 2007.

PEREIRA, R.S.; SANTOS, C.E.; LIRA JUNIOR, M.A.; STANFORD, N.P. Effectiveness of strains selected for cowpea in soil from the semi-arid region of the sertâo of Paraiba. **Revista Bras. de Ciênc. Agrârias**. 3: 105-110, 2008.

PERIN, A., ARAÙJO, A.P.; TEIXEIRA, M.G. Effect of seed size on biomass and nutrient accumulation and on bean productivity. **Pesquisa Agropecuâria Brasileira**, Brasilia, v. 37, n. 12, p. 1711-1718, dec. 2002.

RUMJANEK, N. G.; MARTINS, L. M. V.; XAVIER, G. R.; NEVES, M. C. P. A. Biological nitrogen fixation. In: FREIRE FILHO, F. R.; LIMA, J. A. A.; RIBEIRO, V. Q. (Ed.) **Feijâo-caupi**: technological advances. Teresina: Embrapa Meio-Norte, 2005. p. 280-335.

SANTOS, C. E. R. S.; STAMFORD, N.P., FREITAS, A.D.S.; VIEIRA, I.M.M.B.; SOUTO, S.M.; NEVES, M.C.P; RUMJANEK, N.G. Effectiveness of rhizobia isolated from soils of the Northeast region of Brazil on N2 fixation in peanut (*Arachis hypogaea* L.). **Acta Sci. Agronomy**, 27: 301-307, 2005.

SILVA, R.P.; SANTOS, C.E.R.S.; STAMFORD, N.P.; LIRA JÙNIOR, M.; FREITAS, A.D.S.; RUMJANEK, N.G., XAVIER, G.R. Inoculation with rhizobium in cowpea in the sertao of paraiba. In: CONGRESSO NACIONAL DE FEIJÂO-CAUPI, 1., 2006, Teresina. **Proceedings**... Teresina: Embrapa, 2006. Available at < http://www.cpamn.embrapa.br/anaisconac2006> Accessed on January 26, 2007.

SILVA, W.R.; MARCOS-FILHO, J. Influence of maize seed weight and size on field performance. **Pesquisa Agropecuâria Brasileira**, Brasilia, v.17, n.5, p.1743-1750, 1982.

SOARES, A.L.; PEREIRA, J.P.A.R.; FERREIRA, P.A.A.F.; VALE, H.M.M.; LIMA, A.S.; ANDRADE, M.J.B.; MOREIRA, M.S.. Agronomic efficiency of selected rhizobia and diversity of native nodulating populations in Perdoes (MG). I - caupi. **Revista Brasileira de Ciência do Solo**, v. 30, p.795-802, 2006.

VIEIRA, R.F.; VIEIRA, C.; CALDAS, M.T. Behavior of cowpeas in spring-summer in the Zona da Mata of Minas Gerais. **Pesquisa Agropecuâria Brasileira**, Brasilia, v.35, n.7, p.1359-1365, jul. 2000.

WHITE, J. W.; GONZALEZ, A. Characterization of the negative association between seed yield and seed size among genotypes of common bean. **Field Crops Research**, Amsterdam, v. 23, p. 159-175, 1990.

ZILLI, J. L. E. **Characterization and selection of rhizobium strains for inoculation of caupi [*Vigna unguiculata* (L.) Walp.] in cerrado areas**. 2001. 137 f.

Dissertation (Master's Degree in Soil Science) - UFRRJ, Seropédica.

ZILLI, J.E.; VALICHESKIR.R.; RUMJANEK, N.G.; SIMÔES-ARAÙJO, J.L.; FREIRE FILHO, F.R. E NEVES, M.C.P. Symbiotic efficiency of *Bradyrhizobium* strains isolated from Cerrado soil in caupi. **Pesquisa Agropecuâria Brasileira**, Brasilia, v.41, n.5, p.811-818, May 2006.

CHAPTER III. EFFICIENCY OF RHIZOBIUM STRAINS IN NUTRIENT ABSORPTION IN COWPEAS

3.1 Introduction

The cowpea [*Vigna unguiculata* (L.) Walp.], also known as feijao macassar or feijão de-corda, is of great socio-economic importance to the North and Northeast regions of Brazil, especially for people with lower purchasing power, making it one of the most important components of their diet, since, according to Dantas et al. (2002), its seeds have a high biological food value due to their high protein content, and it is also well adapted to climatic and edaphic adversities due to its hardiness and precocity.

In addition to these characteristics, this legume is able to associate easily with a group of rhizobia present in the soil, designated by Rumjanek et al. (2005) as the miscellaneous cowpea group or tropical rhizobia, a characteristic also found in several other legumes found in the tropics.

In this symbiotic association between legumes and bacteria of the genus *Rhizobium*, specialized organs (nodules) are formed in which the bacteria are able to convert atmospheric nitrogen (N) into ammonia (NH_3), the form used by the plants.

Biological nitrogen fixation (BNF), like other biological processes, is directly influenced by abiotic factors such as temperature, humidity, the presence of gases such as CO_2 and O_2, the concentration of mineral nitrogen in the soil, the presence of phosphorus, acidity, the presence of toxic ions such as Al^3 + and Mn^2 + and the presence of assimilable molybdenum in the soil. The texture of the soil also contributes to the successful survival of rhizobia in the soil and their potential to fix N2 (Mahler and Wollum, 1980).

In addition to nitrogen fixation, inoculation can bring other benefits such as biological control (Pleban et al., 1995), and the production of metabolites that can increase the concentrations of N, P, Fe, Ca and Cu in the aerial part of the host (Cattelan, 1995).

Allied to this, cowpeas are considered an option as a source of organic matter, due to their ability to develop satisfactorily in low-fertility soils and their hardiness. In this way, it is used as a green fertilizer in the recovery of soils that are naturally poor in fertility, or depleted by intensive use, which is very common in the Northeast (Oliveira and Carvalho, 1988).

In order to produce new inoculants, it is necessary to select new strains that are capable of fixing as much atmospheric nitrogen as possible. However, Soares et al. (2006a) comment that the strains selected in the laboratory and greenhouse may not reach their maximum fixation potential in the field, due to, among other factors, competition with the native population established in the soil and adaptation to local environmental conditions.

Thus, the objectives of this study were to evaluate the efficiency of rhizobium strains in nutrient absorption and their possible indication as official inoculants for cowpeas.

3.2 Material and Methods

This work was carried out in the experimental area of the Assis Chateaubriand Agricultural College of the State University of Paraiba (UEPB), located in the municipality of Lagoa Seca - PB, in accordance with the recommendations of the Network of Laboratories for the Recommendation, Standardization and Dissemination of Technology for Microbial Inoculants of Agricultural Interest (RELARE), from August to November 2005. Figure 2.1 shows the monthly rainfall data for the experimental area, obtained from the College's meteorology department.

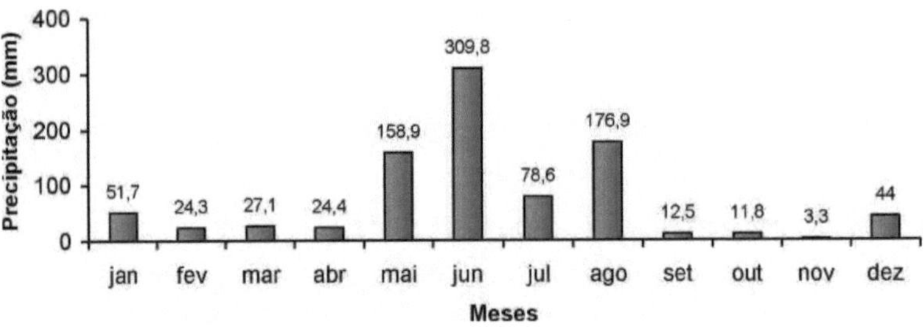

Figure 2.1 Monthly rainfall at the experimental unit of the Colégio Agricola Assis Chateaubriand (2005).

The soil in the experimental area was classified as a Regolithic Neosol whose chemical analysis for fertility purposes, of samples taken between 0.0 and 0.2 m, showed the results in Table 2.1.

Table 2.1 Chemical properties of the soil from the experimental unit at Colégio Assis Chateaubriand (2005).

pH	MO	P	B	Cu	Fe	Mn	Zn
($CaCl_2$)	(g/dm)³		-		(mg/dm	)³ -	
4,5	17,0	20,0	0,3	0,5	62,0	5,1	0,9

K	Ca	Mg	Al	H+Al	SB	T	V
			-(mmolc/ dm³)-		-		(%)
1,8	6,0	2,0	2,0	22,0	9,8	31,8	31,0

SB- sum of bases; T- cation exchange capacity. Deptº of Soils and Plant Nutrition, Esalq.

The soil was prepared by manual weeding to eliminate weeds and harrowing to even out the soil and facilitate fertilization and the opening of the holes. After this phase, the soil was limed to correct the pH and all the plots were fertilized with 80 kg ha^{-1} of P2O5 (simple superphosphate), 30 kg ha^{-1} of K_2O (KCl) and 10 kg ha^{-1} of micronutrients in the form of H_3BO_3, $CuSO_4$, $ZnSO_4$ and $MnSO_4$, respectively.

The experimental plots were 4 m x 6 m and consisted of twelve 4 m rows each. The holes were dug manually with a spacing of 0.5 m x 0.5 m. A useful area of 6 m was considered for each plot^2 , which consisted of six central rows, leaving 1 m on each side.

Five seeds per hole were distributed at sowing, from lots obtained by farmers in the region, following RELARE's recommendations for this project. Thinning was carried out one week after seed germination, leaving 3 (three) plants per hole and a total of 72 (seventy-two) plants in the useful area of each plot.

The seeds of the Corujinha cultivar were inoculated with the rhizobium strains using 30 g of rhizobium mixed with a solution made up of 20 mL of water and 30 g of sugar for every 600 g of seeds, in order to ensure that the inoculant adhered better to the seeds. The seeds were then mixed by hand and dried in the shade for 30 minutes before sowing.

The rhizobium strains BR 3301 (INPA 03-11B or SEMIA 6463) and BR 3302 (UFLA 3-84 or SEMIA 6461) from the culture collection of the Soil Microbiology Laboratory at the Federal University of Lavras; BR 3267 (SEMIA 6462), BR 3299 and BR 3262 from the culture collection of Embrapa Agrobiologia were used to inoculate these seeds. It should be noted that two of the strains mentioned above (BR 3299 and BR 3262) are not yet part of the collection of strains officially recommended for cowpea inoculation and, in this work, their development will be evaluated and compared with that of the other strains currently on the market for inoculants for this crop.

For comparison purposes with the inoculation of the strains, two doses of nitrogen were used: 50 kg ha^{-1} at planting and 80 kg ha^{-1} of nitrogen (40 kg ha^{-1} at planting and 40 kg ha^{-1} at 35 days after sowing) using urea as the nitrogen source; as well as a control treatment (no inoculation and no mineral nitrogen).

Weeding was carried out whenever necessary in order to keep the area clean throughout the experimental period.

The Kjedahl method was used to determine the nitrogen content; potassium was determined by flame photometry and phosphorus by colorimetry, using the metavanadate method, as described by Malavolta et al. (1989).

The following variables were analyzed to assess the effect of the treatments: Percentage of nitrogen in the aerial part, Total nitrogen in the aerial part, Percentage of nitrogen in the seed, Total nitrogen in the seed, Percentage of nitrogen in the nodule, Total nitrogen in the nodule, Percentage of phosphorus in the aerial part, Total phosphorus in the aerial part, Percentage of phosphorus in the seed, Total phosphorus in the seed, Percentage of phosphorus in the nodule, Total phosphorus in the nodule, Percentage of potassium in the aerial part, Total potassium in the aerial part, Percentage of potassium in the grain, Total potassium in the grain, Percentage of potassium in the nodule and Total potassium in the nodule.

The experimental design used was a randomized block design with four replications. The treatments consisted of five rhizobium strains (BR 3301, BR 3302, BR 3267, BR 3299 and BR 3262), mineral nitrogen fertilization (50 kg ha^{-1} and 80 kg ha^{-1}) and a control treatment. The statistical analysis program SISVAR, version 4.6, was used to analyze the results. The means were compared using the Scott-Knott test at 5% probability.

3.3 Results and Discussion

The values for the percentages and contents of nitrogen accumulated in the aerial part, seeds and nodules of the cowpea plants are shown in Table 2.2, and the summary of the analysis of variance for these variables can be found in the Appendix (A 2).

Table 2.2 Average values for the percentage of nitrogen in the aerial part (PNA), nitrogen content in the aerial part (CNA), percentage of nitrogen in the seed (PNS), nitrogen content in the seed (CNS), percentage of nitrogen in the nodule (PNN) and nitrogen content in the nodule (CNN) of cowpea plants subjected or not to inoculation and nitrogen fertilization.

Treatment	**PNA** (%)	**NAC** (mg/plant)	**PNS** (%)	**CNS** (mg/plant)	**NNP** (%)	**CNN** (mg/plant)
BR 3301	4,28 a	54,09 a	3,70 a	184,23 a	5,68 a	4,27 a
BR 3302	4,18 a	50,86 a	3,56 a	111,78 a	5,50 a	4,24 a
BR 3262	4,14 a	45,77 a	3,80 a	167,94 a	4,72 b	4,01 a
BR 3267	4,11 a	58,56 a	3,52 a	147,33 a	5,93 a	5,29 a
BR 3299	3,83 a	42,97 a	3,79 a	166,17 a	5,71 a	2,92 a
80 kg ha^{-1}	4,18 a	57,91 a	3,79 a	201,75 a	5,48 a	3,57 a
50 kg ha^{-1}	4,26 a	46,86 a	4,05 a	267,75 a	4,86 b	1,92 b
Control	3,82 a	33,23 a	3,78 a	147,68 a	3,95 b	0,88 b

CV (%)	8,87	46,72	7,19	51,82	11,36	43,27

Means followed by the same letter do not differ according to the Scott-Knott test at 5%.

The percentage of nitrogen in the aerial part of the plants showed no significant difference when they were subjected to the different treatments (Table 2.2).

Although there were no significant differences between the rhizobium strains, the percentages of nitrogen obtained through leaf diagnosis remained above the level considered critical for the crop, which according to Ambrosano et al. (1996) is 3% in the leaves of the bean. Even in the control treatment, represented here by the native strains, the foliar nitrogen percentages were above this critical value.

Soares et al. (2006a) also used various rhizobium strains (BR 2001, UFLA 03-36, UFLA 03-129, UFLA 03-84 and INPA 03-11B) in Perdoes-MG and found that they did not differ from the treatments with nitrogen fertilization (70 kg ha^{-1}) or without fertilization (control), when assessing the percentage of nitrogen in the aerial part of the cowpea plants.

On the other hand, using the same strains mentioned above, Pereira et al. (2006) found, in the municipality of Iguatama-MG, that the strains UFLA 03-84 and INPA 03-11B provided increases to the aerial part of the cowpea plants similar to those obtained by nitrogen fertilization (80 kg ha^{-1}) and significantly higher than the other strains.

As for the nitrogen content in the aerial part, it can be seen in absolute numbers that the lowest value was achieved by the control treatment, although it did not differ from the others.

Similarly, Ferreira et al. (2000) found that the inoculation of different rhizobium strains and nitrogen fertilization treatments showed similar responses to each other in the accumulation of nitrogen in the leaves, which differs from the results found by other authors (Zilli et al., 2006 and Soares et al., 2006b) who obtained significant increases in nitrogen in the aerial part of cowpea when using inoculants (BR 3262, BR 3267, INPA 03-11B) as nitrogen sources.

The treatments used on the crop also did not produce significant differences in the percentage and accumulation of nitrogen by the seeds.

Analyzing the effects of inoculation on the percentage of nitrogen in cowpea seeds, Pereira et al. (2006) observed that the strains UFLA 03-84 and INPA 03-11B provided significant increases, similar to those achieved with the dose of 80 kg ha^{-1} of nitrogen and superior to the other treatments (without nitrogen, BR 2001, UFLA 03-36 and UFLA 03-129). These authors also found that the INPA 03-11B strain provided significantly higher increases than those found in seeds from plants inoculated with the other strains and the control, but similar to the treatment fertilized with mineral nitrogen (80 kg ha).$^{-1}$

Comparing the averages achieved here by strains that are not yet officially recommended (BR 3362 and BR 3399) with those obtained with inoculants recommended and tested in other studies (Soares et al., 2006b; Pereira et al., 2006), it can be seen that their values do not differ statistically from each other. In this way, the results of this work confirm the tendency of strains BR 3362 and BR 3399 to provide satisfactory percentages and accumulations of nitrogen in cowpea seeds.

As for the results of the percentage of nitrogen in the nodules, there was a significant difference between the treatments. In this variable, the lowest percentages were observed when strain BR 3362, nitrogen fertilization (50 kg ha^{-1}) and the control treatment (native strains) were used.

There were also significant differences in the accumulation of nitrogen in the nodules, but the lowest values were found with the use of mineral fertilizer (50 kg ha^{-1}) and in the presence of the native strains.

Looking at the results, it can be said that the absorption of nitrogen by the plants inoculated with the strains not yet recommended as commercial inoculants (BR 3262 and BR 3299) was similar to that obtained by the strains officially recommended (Diàrio Oficial da Uniâo, 2006) and used in this work.

However, when looking at the rainfall data collected during the course of this study (Figure 2.1), we can see that there was a water deficit for most of the crop's vegetative period. This factor probably contributed significantly to the lack of response of this crop to the treatments since Silveira et al. (2001) found that nitrate assimilation in the leaves and nodule activity were severely affected by the effect of water deficit in cowpea plants previously inoculated with *Bradyrhizobium* spp.

As was the case with the percentage and accumulation of nitrogen in the aerial part and in the seeds (Table 2.2), the same behavior was observed for the percentage of phosphorus in the aerial part, the content of phosphorus in the aerial part, the percentage of phosphorus in the seed and the content of phosphorus in the seed of the cowpea plants, i.e. they were not influenced by the treatments (Table 2.3).

Table 2.3 Average values for percentage of phosphorus in the aerial part (PPA), content of phosphorus in the aerial part (CPA), percentage of phosphorus in the seed (PPS), content of phosphorus in the seed (CPS), percentage of phosphorus in the nodule (PPN) and content of phosphorus in the nodule (CPN) of cowpea plants submitted or not to inoculation and nitrogen fertilization.

Source of N	**PPA** (%)	**CPA** (mg/plant)	**PPS** (%)	**CPS** (mg/plant)	**PPN** (%)	**CPN** (mg /plant)
BR 3301	1,12 a	14,23 a	1,01 a	51,53 a	1,10 a	0,82 a
BR 3302	1,24 a	15,16 a	1,06 a	32,43 a	1,00 a	0,79 a
BR 3262	1,27 a	14,08 a	1,00 a	48,38 a	0,91 b	0,77 a
BR 3267	1,11 a	16,10 a	0,96 a	43,12 a	1,05 a	0,94 a
BR 3299	1,09 a	12,03 a	1,03 a	44,46 a	1,01 a	0,56 a
80 kg ha^{-1}	1,24 a	16,20 a	1,07 a	56,98 a	1,10 a	0,69 a
50 kg ha^{-1}	1,09 a	11,65 a	1,07 a	66,77 a	0,93 b	0,36 b
Control	1,26 a	10,64 a	1,10 a	43,09 a	0,85 b	0,19 b
CV (%)	14,89	41,26	10,64	48,80	9,39	43,08

Averages followed by the same letter do not differ according to the Scott-Knott test at 5%.

Silva et al. (2006) also found no significant differences in the concentration of phosphorus in the aerial part of cowpea plants when they used *Bradyrhizobium* sp. strains (BR 2001 and NFB-700) under different inoculation methods (in the soil or in the seed).

On the other hand, Chabot et al. (1998) found that, in addition to the ability to fix nitrogen, some rhizobium isolates are able to solubilize poorly soluble phosphates and thus make phosphorus available to plants and to themselves.

In addition to rhizobia, some species of bacteria from the genus *Pseudomonas* and *Bacillus* are among the most efficient microorganisms in solubilizing poorly soluble inorganic phosphates in the soil (Rodriguez and Fraga, 1999), and it was precisely with a strain of *Bacillus subtilis* that Gaing and Gaur (1991), working with beans in phosphorus-deficient soil, observed that this bacterium increased the absorption of phosphorus, nitrogen, grain production and biomass of the crop studied.

Despite the practical and economic importance of the legume-rhizobium association, few studies have been carried out, mainly on the chemical factors of the effect of this inoculation on the percentages of phosphorus and even many other elements involved in plant growth and development.

Other determinations regarding the percentage and content of phosphorus were made in the nodules of the cowpea plants and, in both variables, significant differences were found when subjected to the treatments used (Table 2.3).

The percentages of phosphorus in the nodules were lower when strain BR 3262, mineral nitrogen fertilization (50 kg ha^{-1}) and the control treatment (native strains) were used.

In the case of the accumulated phosphorus content in this plant organ, the lowest amounts were found when 50 kg of N ha was used^{-1} and in the control treatment.

In plants of *Piptadenia gonoacantha* (an arboreal legume) inoculated with arbuscular mycorrhizal fungi (AMFs), Jesus et al. (2005) observed a higher phosphorus content in the aerial part than those not inoculated, showing the beneficial effect of the fungus in supplying phosphorus to the plants.

Comparing the results of the percentage and content in the nodules in relation to nitrogen (Table 2.2) and phosphorus (Table 2.3), it can be seen that the lowest concentrations of both nitrogen and phosphorus were obtained with the same treatments (50 kg ha^{-1} and control). This indicates that the strains that are efficient and/or inefficient at absorbing nitrogen are also efficient in relation to phosphorus.

According to Silva et al. (2006), plants dependent on biological nitrogen fixation require more phosphate than plants that use mineral N exclusively. Other researchers (Siqueira and Franco, 1988) add that low levels of phosphorus can affect the symbiosis by reducing the supply of photosynthate to the nodule, reducing the bacterial growth rate and the total population of *Rhizobium* and *Bradyrhizobium*.

In jacatupé plants (*Pachyrhizus erosus* (L.) Urban), Stamford et al. (1999) found a positive response to phosphorus levels in the amount of total nitrogen accumulated. These results corroborate those found in the same crop by Figueiredo et al. (1996) and by Carvalho et al. (1988) in *Centrosema* and *Stilosanthes,* who observed increases in nitrogen in these crops when the amount of phosphorus available to the plants was increased.

Phosphorus, which is deficient in most tropical soils, is a nutrient that has a marked effect on nitrogenase activity, due to the high energy expenditure promoted by biological nitrogen fixation (Straliotto and Rumjanek, 1999).

Several experiments have been carried out in the field and greenhouse to evaluate the effect of different levels of this nutrient on symbiotic efficiency and plant productivity under symbiotic conditions (Pereira and Bliss, 1987, 1989), and there has always been a positive response to phosphorus fertilization.

Given these observations, it is likely that the availability of phosphorus also influenced the results obtained in the parameters evaluated in Chapter 1, as well as the amounts of nitrogen absorbed (Table 2.2).

This availability of phosphorus may have been interrupted due to the drought, since periods of veranico lasting approximately ten days cause the flow of inorganic phosphate (Pi) from the soil to the plant to be practically paralyzed, which can cause significant losses in productivity (Mouat and Nes, 1986; Novais and Smyth, 1999).

The results of the potassium concentrations found in the aerial part, seeds and pods of the cowpea can be seen in Table 2.4. With regard to the percentage of potassium found in the aerial part of the cowpea plants, it can be seen that only the treatments fertilized with mineral nitrogen (50 and 80 kg.ha^{-1}) saw significant increases in this element.

On the other hand, no significant differences were found for the potassium content in the aerial part of the cowpea plants, although the lowest absolute value was recorded with the native strains.

When comparing the concentrations of phosphorus (Table 2.3) with those of potassium (Table 2.4) in the aerial part of the cowpea plants, there were higher concentrations of the latter element, in agreement with Costa et al. (2003) when they verified the high capacity of cowpea to accumulate this ion in its leaves.

In relation to the accumulated nitrogen content in the aerial part of jacatupè, Stamford et al. (1999) found that there was no significant difference between the treatments with and without the addition of potassium (K_2O) and magnesium (MgO) fertilizer.

Table 2.4 Average values of the percentage of potassium in the aerial part (PKA), potassium content in the aerial part (CKA), percentage of potassium in the seed (PKS), potassium content in the seed (CKS), percentage of potassium in the nodule (PKN) and potassium content in the nodule (CKN) of cowpea plants subjected or not to inoculation and nitrogen fertilization.

Source of N	PKA (%)	CKA (mg/plant)	PKS (%)	CKS (mg/plant)	PKN (%)	CKN (mg /plant)
BR 3301	2,20 b	27,68 a	1,60 a	84,80 a	1,74 a	1,31 a
BR 3302	2,40 b	28,75 a	1,60 a	48,51 a	1,49 a	1,19 a
BR 3262	2,82 b	31,22 a	1,60 a	70,58 a	1,67 a	1,40 a
BR 3267	2,46 b	36,59 a	1,53 a	71,33 a	1,53 a	1,41 a
BR 3299	2,00 b	27,75 a	1,52 a	60,29 a	1,57 a	0,82 b
80 kg ha^{-1}	3,29 a	44,41 a	1,53 a	89,44 a	1,77 a	1,14 a
50 kg ha^{-1}	3,82 a	43,93 a	1,56 a	106,53 a	1,68 a	0,63 b
Control	2,71 b	23,50 a	1,57 a	64,39 a	1,47 a	0,33 b
CV (%)	18,76	52,29	8,97	51,92	15,17	44,92

Averages followed by the same letter do not differ according to the Scott-Knott test at 5%.

However, in the cowpea crop, Stamford et al. (1980) found that the application of high doses of potassium reduced the amount of total nitrogen accumulated in the plant.

Both the percentage and the potassium content of the seeds were not affected by the treatments, however, the potassium content in the nodules varied significantly, with lower amounts of this element being observed with the control treatment, 50 kg of N ha^{-1} and with strain BR 3299, respectively.

The concentrations of potassium (Table 2.4) were higher than those of phosphorus (Table 2.3), indicating the importance of this element for cowpeas and proposing further research into the effect of inoculation on the availability of this element for cowpeas.

3.4 Conclusions

The absorption of nitrogen, phosphorus and potassium in plants inoculated with *Bradyrhizobium* strains was similar to that achieved by plants fertilized with mineral nitrogen (urea) at doses of 50 and 80 kg ha^{-1} and in the control treatment;

The responses of the control treatment indicate a native population of rhizobium strains;

The results indicate a similar development between the strains not yet officially recommended (BR 3262 and BR 3299) and those already indicated as official inoculants for cowpeas.

3.5 Bibliographical references

AMBROSANO, E.J.; TANAKA, R.T.; MASCARENHAS, H.A.A.; RAIJ. B. van; QUAGGIO, J.A.; CANTARELLA, H. Legumes and oilseeds. In: RAIJ, B. van; CANTARELLA, H.; QUAGGIO, J.A.; FURLANI A.M.C. (Ed.) **Recomendações de adubaçâo e calagem para o Estado de Sâo Paulo**. 2.ed. Campinas: Instituto Agronômico/Fundaçâo IAC, 1996. chap.19, p.187-199.

CARVALHO, M.M.; SARAIVA,O.F.; OLIVEIRA, F.T.T.; MARTINS, C.E. Responses of tropical forage legumes to liming and phosphorus, in a greenhouse. **Revista Brasileira de Ciência do Solo**, Viçosa, v.12, p.153-159, 1988.

CATTELAN, A. J. Increase in the number of root hairs in soybean seedlings inoculated with growth-promoting bacteria. p. 393-397. In: **Brazilian Symposium on Soil Microbiology**, 3. Embrapa, Londrina. 421 p. **Proceedings**, 1995.

CHABOT, R.; BEAUCHAMP, C.J.; KLOEPPER, J.W.; AUTON, H. Effect of phosphorus on root colonization and growth promotion of maize by bioluminescent mutants of phosphate-solubilizing *Rhizobium leguminosarum* bv *phaseoli*. **Soil Biology Biochemistry**, Oxford, v. 30, p. 1615-1618, 1998.

COSTA, P.H.; SILVA, J.V.; BEZERRA, M.A.; ENÉAS FILHO, J.; PRISCO, J.T.;

GOMES FILHO, E. Growth and levels of organic and inorganic solutes in cultivars of *Vigna unguiculata* submitted to salinity. **Revista Brasileira de Botànica**, Sao Paulo, v.26, n.3, p.289-297, 2003.

DANTAS, J.P.; MARINHO, F.J.L.; FERREIRA, M.M.; AMORIM, M.S.N.; ANDRADE, S.I.O.; SALES, A.L. Evaluation of caupi genotypes under salinity. **Revista Brasileira de Engenharia Agricola e Ambiental**, Campina Grande, v.6, n.3, p.425430, 2002.

OFFICIAL GAZETTE OF THE UNION. **Instruçâo normativa no 10**, de 21 de março de 2006, seção 1. Available at < http://www.diario-oficial.com.br/servicos/diario_eletronico/index.asp> Accessed on January 30, 2007.

FERREIRA, A.N.; ARF1, O.; CARVALHO, M.A.C.; ARAÙJO, R.S.; SA, M.E.; BUZETTI, S. Strains of rhizobium tropici in bean inoculation. **Scientia Agricola**, Piracicaba, v.57, n.3, p.507-512, 2000.

FIGUEIREDO, M.V.B.; MEDEIROS; R.; STAMFORD, N.P.; SANTOS, C.E.R.S. Effect of fertilization with different potassium/magnesium ratios on jacatupé in Yellow Latosol with and without inoculation with *Bradyrhizobium* sp. **Revista Brasileira de Ciência do Solo**, Campinas, v.20, p.49-54, 1996.

GAING, S. & A C. GAUR. Thermotolerant phosphate solubilizing microorganisms and their interaction with mung bean. **Plant and Soil**, v. 133, p. 141-149, 1991.

JESUS, E.C.; SCHIAVO, J.A.; FARIA, S.M. Dependence on mycorrhizae for the nodulation of tropical leguminous trees. **Revista Arvore**, Viçosa-MG, v.29, n.4, p.545-552, 2005.

MAHLER, R.L., WOLLUM, A.C. Influence of water potential on the survival of rhizobia in agoldsboro loamy sand. **Soil Science Society of American Journal**. Madison, n.44, v.3, p.988-992, 1980.

MALAVOLTA, E.; VITTI, G.C.; OLIVEIRA, S.A. de. **Evaluation of plant nutritional status** - principles and applications. Piracicaba: POTAFOS, 1989. 201p.

MOUAT, M.C.H.; NES, P. Influence of soil water content on the supply of phosphate to plants. **Australian Journal Soil Research**, Canberra, v. 24, p. 435-440, 1986.

NOVAIS, R.F.; SMYTH, T.J. **Soil and plant phosphorus conditions in tropical environments**. Viçosa: Editora UFV, 1999, p.1-7.

OLIVEIRA, I.P.; CARVALHO, A.M. The caupi crop in the climate and soil conditions of the humid tropics of semi-arid Brazil. In: ARAÙ- JO, J.P.P.; WATT, E.E. org. **Caupi in Brazil**.

Brasilia: IITA/EMBRAPA, 1988. p. 63-96.

PEREIRA, J. P. A. R.; FERREIRA, P. A. A.; VALE, H. M. M.; NOGUEIRA, C. O. G.; SOARES, A. L. L.; MOREIRA, F. M. S.; ANDRADE, M. J. B. Nodulation and productivity of cowpea cv. Poços de Caldas by selected rhizobium strains in Iguatama-MG. In: CONGRESSO NACIONAL DE FEIJÂO-CAUPI, 1., 2006, Teresina. **Proceedings**... Teresina: Embrapa, 2006. Available at < http://www.cpamn.embrapa.br/anaisconac2006> Accessed on January 26, 2007.

PEREIRA, P.A.A.; BLISS, F.A. Nitrogen fixation and plant growth of common bean (*Phaseolus vulgaris* L.) at different levels of phosphorus availability. **Plant and Soil**, Dordrecht, v.104, p.79-84, 1987.

PEREIRA, P.A.A.; BLISS, F.A. Selection of common bean (*Phaseolus vulgaris* L.) for N2 fixation at different levels of available phosphorus under field and environmentally controlled conditions. **Plant and Soil**, Dordrecht, v.115, p.75-82, 1989.

RODRIGUEZ, H.; FRAGA, R. Phosphate solubilizing bacteria and their role in plant growth promotion. **Biotechnology Advances**, v. 17, p. 319-339. 1999.

RUMJANEK, N. G.; MARTINS, L. M. V.; XAVIER,G. R.; NEVES, M. C. P. Biological Nitrogen Fixation. In: FREIRE FILHO, F. R.; LIMA, J. A. A.; SILVA, P. H. S.; VIANA, F. M. P. (Org.). **Feijâo-caupi**: avanços tecnológicos. p. 281-335, 2005.

SILVA, V.N.; SILVA, L.E.S.F.; FIGUEIREDO, M.V.B. Co-inoculation of caupi seeds with *bradyrhizobium* and *paenibacillus* and their efficiency in the absorption of calcium, iron and phosphorus by the plant. **Pesquisa Agropecuâria Tropical**, Goiânia, v. 36, n. 2, p. 95-99, 2006.

SILVEIRA, J.A.G.; COSTA, R.C.L.; OLIVEIRA, J.T.A. Drought-induced effects and recovery of nitrate assimilation and nodule activity in cowpea plants inoculated with *bradyrhizobium* spp. Under moderate nitrate level. **Brazilian Journal of Microbiology,** Sâo Paulo, v. 32, p.187-194, 2001.

SIQUEIRA, J. O. & A. A. FRANCO. **Soil biotechnology**: fundamentals and perspectives. Nagy, Brasilia. 235 p. 1988.

SOARES, A. L. L.; MOREIRA, F. M. S.; ANDRADE, M. J. B. Nodulation and productivity of cowpea cv. Poços de Caldas by selected rhizobium strains in Iguatama-MG. In: CONGRESSO NACIONAL DE FEIJAO-CAUPI, 1., 2006, Teresina. **Proceedings**... Teresina: Embrapa, 2006 a. Available at < http://www.cpamn.embrapa.br/anaisconac2006>

Accessed on January 26, 2007.

SOARES, A.L.L.; PEREIRA, J.P.A.R.; FERREIRA, P.A.A.F.; VALE, H.M.M.; LIMA, A.S.; ANDRADE, M.J.B.; MOREIRA, M.S.. Agronomic efficiency of selected rhizobia and diversity of native nodulating populations in Perdoes (MG). I - caupi. **Revista Brasileira de Ciência do Solo**, Viçosa, v. 30, p.795-802, 2006 b.

STAMFORD, N.P.; NEPTUNE, A.M.L.; SILVA, I.P. Effect of potassium in the presence of N-mineral on nodulation, growth and nutrient uptake by (*Vigna unguiculata* (L.) Walp.). **Revista Brasileira Ciência do Solo**, Campinas, v.4, p.99103, 1980.

STAMFORD, N.P; SANTOS, C.E.R.S; MEDEIROS, R.; FREITAS, A.D.S. Effect of fertilization with phosphorus, potassium and magnesium on jacatupé infected with rhizobium in an alkalic latosol. **Pesquisa Agropecuària Brasileira**, Brasilia, v.34, n.10, p.18311838, 1999.

STRALIOTTO, R; RUMJANEK, N.G. Biodiversity of rhizobium nodulating bean (*phaseolus vulgaris* l.) and the main factors affecting symbiosis. EMBRAPA: **Documento**, n. 94, 1999.

ZILLI, J.E.; VALICHESKIR.R.; RUMJANEK, N.G.; SIMOES-ARAÙJO, J.L.; FREIRE FILHO, F.R. E NEVES, M.C.P. Symbiotic efficiency of *Bradyrhizobium* strains isolated from Cerrado soil in caupi. **Pesquisa Agropecuària Brasileira**, Brasilia, v.41, n.5, p.811-818, May 2006.

CHAPTER IV. EFFICIENCY OF RHIZOBIUM STRAINS IN THE PRODUCTION AND QUALITY OF COWPEA SEEDS - 2006 CROP YEAR

4.1 Introduction

The cowpea (*Vigna unguiculata* (L.) Walp.) is the most important grain legume in Brazil's Semi-Arid region, and plays the role of supplying part of the protein needs of the region's poorest populations (Teixeira et al., 1988).

Traditionally grown under rainfed conditions, cowpeas have emerged as an option for irrigated cultivation in the Northeast. In these areas, it is grown in succession to another crop of greater economic value, in order to take advantage of the residual effect of fertilization and a smaller supply of the product in some off-season periods (Santos et al., 2000). Medeiros et al. (2005) also add that this crop is a viable alternative due to its significant socio-economic importance in the region, as it is the main source of low-cost vegetable protein for human consumption.

According to Yokoyama et al. (2000), of all bean production in the country, around 77% comes from the *Phaseolus* genus and 23% from the *Vigna* genus, with the Northeast region being the main producer in terms of planted area and production, while around 60% of all production comes from the *Vigna* genus.

Despite its socio-economic importance for this region, over the years this crop has shown low productivity, around 400 kg ha^{-1} , largely due to the growing conditions, low soil fertility and, most of the time, without the adoption of advanced technologies.

One of the alternatives that is becoming increasingly studied and used by researchers and farmers is the use of atmospheric nitrogen-fixing bacteria, with the aim of supplying a large part of the nitrogen needed for the development and production of these plants. However, for this to happen, the legume must be efficiently nodulated by the specific rhizobium.

Nitrogen is the most abundant component in the atmosphere and is found in a combined form (N_2) that plants are unable to use. In terrestrial systems, nitrogen is the most limiting chemical element for plant growth. However, atmospheric nitrogen can be utilized through a natural process known as biological nitrogen fixation (BNF). In this process, the triple bond of N2 is broken by diazotrophic microorganisms and the molecular nitrogen is converted into ammonia and made available to plants (Pedrosa et al., 2000).

This process is an ecological and economical alternative to nitrogen fertilizers because, in addition to dispensing with the use of these fertilizers, all biologically fixed nitrogen is used by the plant when in association (Franco and Dobereiner, 1994).

These bacteria form nodules on the roots and sometimes on the stems of various host plants and the species of these bacteria were initially classified within a single genus (*Rhizobium*) and then reclassified within the *Rhizobiaceae* family into *Rhizobium, Bradyrhizobium, Azorhizobium, Sinorhizobium* and *Mesorhizobium* (Pereira, 2002).

There are currently hundreds of strains distributed within these genera, which leads researchers to select the most suitable for inoculation, also depending on the host species, since some strains show specificity.

In this context, the aim of this work was to evaluate the efficiency of rhizobium strains in the production and physiological quality of cowpea seeds in the second year of planting.

4.2 Material and Methods

This research was carried out in the experimental area of the Assis Chateaubriand Agricultural College - State University of Paraiba (UEPB), located in the municipality of Lagoa Seca - PB, in accordance with the guidelines of the Network of Laboratories for the Recommendation, Standardization and Dissemination of Technology for Microbial Inoculants of Agricultural Interest (RELARE), during the period from June to September 2006. The accumulated monthly rainfall, according to data obtained from the college's meteorology department, is shown in Figure 3.1.

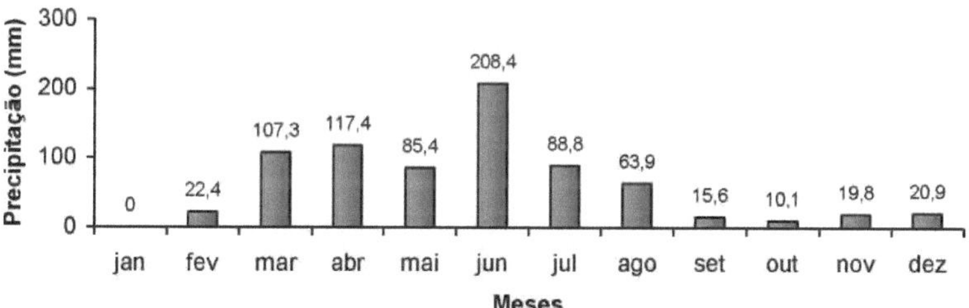

Figure 3.1 Monthly rainfall at the College's experimental unit

Agricola Assis Chateaubriand (2006).

The soil in the experimental area was classified as Neossolo Regolitico, and the results of the chemical analysis for fertility purposes, on a representative sample between 0.0 and 0.2 m, are shown in Table 3.1.

Table 3.1 Chemical properties of the soil at the experimental unit of the Colégio Agricola

Assis Chateaubriand (2006).

pH	MO	P	B	Cu	Fe	Mn	Zn
($CaCl_2$)	(g/dm)³	-	-		(mg/dm	)³ -	
4,9	22,0	22,0	0,44	0,6	52,0	5,1	0,9

K	Ca	Mg	Al	H+Al	SB	T	V
		(mmolc/			dm3) -		(%)
2,2	13,0	5,0	1,0	20,0	21,2	41,2	51,0

58- sum of bases; T- cation exchange capacity. Department of Soils and Plant Nutrition, Esalq. University of São Paulo.

Soil preparation consisted of harrowing to even out the soil and make it easier to open the holes and fertilize, and manual weeding was carried out to eliminate weeds during the experimental phase.

The experimental plots measured 4 m x 6 m, giving a total of twelve 4 m rows each. The pits were spaced 0.5 m x 0.5 m and were opened manually with the aid of hoes. Data was collected for statistical analysis in each plot, in a useful area consisting of six central rows, with 1 m on either side of them, establishing a space of 6 m .²

Following guidance from RELARE for this project, cowpea seeds, cultivar Corujinha, were obtained from farmers in the region and planted so that each hole contained five seeds. The cowpea seeds, cultivar Corujinha, were inoculated with rhizobium strains, using 30 g of rhizobium for every 600 g of seeds, mixed in a solution made up of 20 mL of water and 30 g of sugar, so that the inoculant would adhere better to the seeds. The seeds were then mixed by hand and dried in the shade for 30 minutes before sowing.

In the process of inoculating these seeds, we used the rhizobium strains BR 3301 (INPA 03-11B or SEMIA 6463) and BR 3302 (UFLA 3-84 or SEMIA 6461), from the culture collection of the Soil Microbiology Laboratory at the Federal University of Lavras; BR 3267 (SEMIA 6462), BR 3299 and BR 3262, from the culture collection of Embrapa Agrobiologia. It is worth pointing out that the last two strains mentioned have not yet been officially recommended for cowpea inoculation and, in this work, their development will be evaluated and compared with that of the other strains currently on the market for inoculants for this crop.

For comparison purposes with the inoculation of these strains, the following doses of nitrogen fertilizer (urea) were used: 50 kg ha^{-1} of nitrogen (at planting) and 80 kg ha^{-1} of nitrogen (40 kg ha^{-1} at planting and 40 kg ha^{-1} at 35 days); as well as a control treatment (no inoculation and no mineral N).

Thinning was carried out one week after germination, leaving three (3) plants per hole, giving a total of seventy-two (72) plants in the useful area of each plot. Weeding was carried out whenever necessary in order to keep the area clean throughout the experimental period.

The effect of the treatments was evaluated using the following variables:

a) Number of nodules - obtained by manually counting the nodules contained in the roots of ten plants, collected from the third row on each side of the plot, 35 days after the seedlings emerged;

b) Dry weight of the nodules - the nodules removed to determine the previous variable were dried in a forced air circulation oven at a temperature of 65° C for 48 hours and then weighed, with the result expressed in mg plant ;$^{-1}$

c) Nodule weight - the average weight of each nodule was determined by the ratio between the dry weight of the nodules and the number of nodules, expressed in mg nodule ;$^{-1}$

d) Nitrogen in the pod, aerial part and seed - The methodology described by Malavolta et al. (1989) was used to determine the nitrogen content;

e) Dry weight of the aerial part and root - three plants were collected from the second row of the plot to determine the dry weight of the aerial part (stem and leaves) and the root, at 35 and 68 DAE. Both parts were dried in a forced-air oven at 65° C for 72 hours and then weighed separately;

f) Leaf area - leaf disks were detached from the leaf limb using a punch with a known area. The total leaf area was estimated using the known area of the detached leaf disks, their weight and the weight of the leaf, taken using an analytical balance;

g) Weight of pods per plant - after drying, the pods were weighed and the ratio between their mass and the number of plants collected in the useful area of the plots was calculated;

h) Number of pods per plant - determined from the ratio of the total number of pods in the useful area to the number of plants in this area;

i) Number of seeds per pod - ten pods were taken at random from the useful area and the seeds were counted to determine their average per pod;

j) Pod length - based on the average of the ten pods taken at random to determine the previous variable, where they were measured using a ruler graduated in centimeters;

l) Seed yield - the seed yield of the useful area was determined, corrected to 13% moisture and expressed in kg ha ;$^{-1}$

m) Weight of 100 seeds - determined by weighing eight replicates of 100 seeds per treatment (Brasil, 1992);

n) Seed yield per plant - determined by the ratio between the production of the useful area and the number of plants in it, with the result corrected to 13% moisture;

o) Germination - The germination capacity of the seeds was assessed using the standard germination test, in accordance with the Rules for Seed Analysis (Brazil, 1992);

p) Dry weight of seedlings - carried out at the end of the germination test where the seedlings were placed in paper bags and taken to dry in an oven at 65° C until they reached a constant weight, based on Nakagawa (1999).

q) Emergence of seedlings in sand - conducted in a greenhouse where four replicates of 50 seeds were distributed in plastic trays containing sterilized sand and moistened with a quantity of water equivalent to 60% of its retention capacity (Brasil, 1992);

r) First seedling emergence count - the emergence data of the first seedlings to emerge from the aforementioned test were computed;

s) emergence rate in sand - this evaluation also used the seedling emergence test in sand, i.e. daily evaluations of normal seedlings were carried out from the first count until all seedlings had emerged. The emergence speed index (ESI) was calculated using the formula proposed by Maguire (1962).

A randomized block design with six replications was used, with the treatments represented by rhizobium strains (BR 3301, BR 3302, BR 3267, BR 3299 and BR 3262), nitrogen fertilization (50 kg ha^{-1} , 80 kg ha^{-1}) and a control treatment (control). The data was used using the SISVAR statistical analysis program, version 4.6. The means of the treatments were compared using the Scott-Knott test at 5% probability. A joint analysis was also carried out using the F test to study the similar variables in the two years of cowpea cultivation.

4.3 Results and Discussion

The summary of the analysis of variance for the variables studied in this chapter can be found in the Appendix (A 3).

The data on the number of nodules, nodule dry weight, aerial part and root dry weight and total dry weight show that the treatments used to grow the cowpea had no influence on these variables (Table 3.2). Some of these variables (NuN, PSA, PSR) also showed no significant differences when the cowpea was analyzed in its first year of cultivation (Chapter I).

Table 3.2. Average values of nodule number (NuN), nodule dry weight (MSN), aerial part dry weight (MSA), root dry weight (MSR) and total dry weight (MST) from cowpea plants harvested at 35 A.D., subjected or not to inoculation and nitrogen fertilization.

Treatments	NuN (ud/plant)	PSN (mg/plant)	PSA (g/plant)	PSR (g/plant)	PST (g/plant)
BR 3301	12,65 a	38,95 a	0,95 a	0,38 a	1,33 a
BR 3302	14,75 a	49,05 a	1,27 a	0,42 a	1,69 a
BR 3262	19,33 a	53,18 a	1,20 a	0,39 a	1,59 a
BR 3267	15,98 a	46,63 a	1,28 a	0,40 a	1,68 a
BR 3299	17,56 a	49,93 a	1,18 a	0,38 a	1,56 a
80 kg ha^{-1}	10,31 a	27,70 a	1,05 a	0,27 a	1,33 a
50 kg ha^{-1}	14,51 a	41,05 a	1,22 a	0,34 a	1,56 a
Control	17,86 a	52,35 a	1,26 a	0,42 a	1,68 a
CV (%)	32,10	45,08	25,75	24,35	24,51

Averages followed by the same letter do not differ according to the Scott-Knott test at 5%.

Although there were no significant differences between the strains, it is clear that the presence of native strains at the experimental site is confirmed, since those seeds that were not inoculated also gave rise to nodulated plants.

This nodulation in plants infected by native strains does not always occur, as Lima et al. (2005) found no nodules on control plants with and without mineral nitrogen, unlike what was observed in plants from rhizobia-inoculated seeds.

The symbiotic specificity and wide range of this group of native rhizobia in the soils of tropical regions have been pointed out as hindering the practice of inoculation by making it difficult or even preventing the establishment of more efficient inoculating strains (Singleton et al. 1992; Mpepereki et al., 1996).

The joint analysis of the common variables (NuN, MSN, MSA, MSR) in the two years of cowpea cultivation is shown in Table 3.3.

Table 3.3 Effect of the agricultural year on the variables number of nodules (NN), nodule dry weight (MSN), aerial part dry weight (MAS) and root dry weight (MSR) at 35 A.D.E. of cowpea plants subjected or not to inoculation and nitrogen fertilization.

TREAT	**NN** (ud/plant)	**MSN** (mg/plant)	**MSA** (g/plant)	**MSR** (g/plant)

	YEAR		YEAR		YEAR		YEAR	
	2005	2006	2005	2006	2005	2006	2005	2006
BR 3301	21,5 a	12,6 b	75,1 a	38,9 b	1,25 a	0,95 a	0,30 a	0,38 a
BR 3302	15,7 a	14,7 a	79,0 a	49,0 b	1,19 a	1,27 a	0,31 a	0,42 a
BR 3262	17,7 a	19,3 a	84,3 a	53,1 a	1,10 a	1,20 a	0,36 a	0,39 a
BR 3267	20,5 a	15,9 a	88,4 a	46,6 b	1,38 a	1,28 a	0,40 a	0,40 a
BR 3299	14,7 a	17,5 a	54,4 a	49,9 a	1,09 a	1,18 a	0,34 a	0,38 a
80 kg ha^{-1}	15,7 a	10,3 a	63,4 a	27,7 b	1,34 a	1,05 a	0,32 a	0,27 a
50 kg ha^{-1}	11,5 a	14,5 a	38,9 a	41,0 a	1,06 a	1,22 a	0,33 a	0,34 a
Control	12,0 a	17,8 a	22,3 b	52,3 a	0,86 a	1,26 a	0,26 b	0,42 a

Averages followed by the same letter do not differ on the same line by the F test at 5%.

For the number of nodules, only strain BR 3301 showed a different behavior, with its lowest value in the second year of planting the cowpea.

As for the dry weight of the nodules, most of the strains (BR 3301, BR 3302, BR 3267) and one of the fertilizer treatments (80 kg ha^{-1}) gave the highest values in the first year of cultivation.

According to the data obtained, positive inoculation responses are more easily obtained in areas with low populations of cowpea nodulating rhizobia established in the soil, a fact that tends to occur during periods of prolonged drought (Martins et al., 2003).

Root dry weight was similar for most of the treatments, except for the control treatment, which showed the lowest value in the first year of planting.

Table 3.4 shows the variables relating to nitrogen accumulation in the different vegetative parts of the cowpea. In this second year of planting, no significant differences were found between the treatments used when analyzing the different variables (Table 3.4).

Table 3.4 Average values for the percentage of nitrogen in the aerial part (PNA), nitrogen content in the aerial part (CNA), percentage of nitrogen in the seed (PNS), nitrogen content in the seed (CNS), percentage of nitrogen in the nodule (PNN) and nitrogen content in the nodule (CNN) of cowpea plants subjected or not to inoculation and nitrogen fertilization.

Treatment	**PNA** (%)	**NAC** (mg/plant)	**PNS** (%)	**CNS** (mg/plant)	**PNN** (%)	**CNN** (mg /plant)
BR 3301	3,16 a	29,48 a	3,16 a	405,19 a	4,00 a	1,50 a
BR 3302	3,00 a	38,39 a	3,16 a	420,85 a	3,83 a	1,77 a
BR 3262	3,00 a	35,82 a	3,00 a	351,85 a	3,66 a	2,05 a
BR 3267	3,00 a	40,01 a	3,00 a	477,39 a	4,16 a	1,95 a

BR 3299	3,00 a	35,49 a	3,00 a	415,29 a	3,83 a	1,99 a
80 kg ha^{-1}	3,00 a	31,88 a	3,33 a	371,32 a	3,83 a	0,93 a
50 kg ha^{-1}	2,50 a	29,32 a	3,16 a	402,44 a	4,16 a	1,72 a
Control	2,16 a	27,13 a	3,00 a	393,54 a	3,83 a	1,98 a
CV (%)	21,81	33,92	12,75	41,19	11,95	45,41

Averages followed by the same letter do not differ according to the Scott-Knott test at 5%.

On the other hand, in the first year of cultivation (Chapter II), there were significant differences between the treatments when the percentage of nitrogen in the nodule and the nitrogen content in the nodule were evaluated.

The results of the joint analysis for the variables relating to nitrogen concentrations are shown in Table 3.5.

When comparing the results obtained in the two consecutive years of planting, it can be seen that the nitrogen content of the aerial part, the nitrogen content in the nodule and the total nitrogen in the nodules were lower in the second year of cowpea cultivation. However, the total nitrogen found in the seeds was higher in the last year of cultivation.

This suggests that, in the second year of cultivation, nitrogen was remobilized more from its vegetative parts (aerial parts and roots) to the seeds, probably due to the greater availability of water in the soil when the crop was in full reproductive stage.

The accumulation of nitrogen by cowpea plants as a result of inoculation with rhizobium strains still has some contradictory results, as some studies show the effectiveness of these bacteria in increasing nitrogen concentrations (Zilli et al., 2006; Morgado et al., 2006), while others have found no significant differences in their use (Alcântara et al., 2006; Silva et al., 2006).

Table 3.5 Effect of the agricultural year on the variables percentage of nitrogen in the aerial part (PNA), nitrogen content in the aerial part (CNA), percentage of nitrogen in the seed (PNS), nitrogen content in the seed (CNS), nitrogen percentage in the nodule (PNN) and nitrogen content in the nodule (CNN) of cowpea plants subjected or not to inoculation and nitrogen fertilization.

TREAT	**PNA** (%)		**CNA** (mg /plant)		**PNS** (%)		**CNS** (mg /plant)		**PNN** (%)		**CNN** (mg /plant)	
	YEAR		YEAR		YEAR		YEAR		YEAR		YEAR	
	2005	2006	2005	2006	2005	2006	2005	2006	2005	2006	2005	2006
BR 3301	4,28 a	3,16 b	54,09 a	29,48 b	3,7 a	3,1 b	184,2 b	405,1 a	5,6 a	4,0 b	4,2 a	1,5 b

BR 3302	4,18 a	3,00 b	50,86 a	38,39 a	3,5 a	3,1 a	111,7 b	420,8 a	5,5 a	3,8 b	4,2 a	1,7 b
BR 3262	4,14 a	3,00 b	45,77 a	35,82 a	3,8 a	3,0 b	167,9 b	351,8 a	4,7 a	3,6 b	4,0 a	2,0 b
BR 3267	4,11 a	3,00 b	58,56 a	40,01 a	3,5 a	3,0 a	147,3 b	477,3 a	5,9 a	4,1 b	5,2 a	1,9 b
BR 3299	3,83 a	3,00 b	42,97 a	35,49 a	3,7 a	3,0 b	166,1 b	415,2 a	5,7 a	3,8 b	2,9 a	1,9 a
80 kg ha^{-1}	4,18 a	3,00 b	57,91 a	31,88 b	3,7 a	3,3 b	201,7 a	371,3 a	5,4 a	3,8 b	3,5 a	0,9 b
50 kg ha^{-1}	4,26 a	2,50 b	46,86 a	29,32 a	4,0 a	3,1 b	267,7 b	402,4 a	4,8 a	4,1 a	1,9 a	1,7 a
Control	3,82 a	2,16 b	33,23 a	27,13 a	3,7 a	3,0 b	147,6 b	393,5 a	3,9 a	3,8 a	0,8 a	1,9 a

Averages followed by the same letter in the row do not differ by the F test at 5%.

With regard to the presence of crude protein in the different parts of the cowpea plants (vegetative and reproductive), it can be seen that only the crude protein content of the aerial part suffered a significant effect from the treatments, since those plants grown under nitrogen fertilization (50 kg ha^{-1}) and in the control treatment showed the lowest concentrations of this constituent (Table 3.6).

Table 3.6 Average values for the percentage of crude protein in the aerial part (PPBA), crude protein content in the aerial part (CPBA), percentage of crude protein in the seed (PPBS), total crude protein content in the seed (CPBS), percentage of crude protein in the nodule (PPBN) and crude protein content in the nodule (CPBN) of cowpea plants subjected or not to inoculation and nitrogen fertilization.

Source of N	TPBPA (%)	PBTPA (mg/plant)	TPBS (%)	PBTS (mg/plant)	TPBN (%)	PBTN (mg /plant)
BR 3301	19,50 a	184,24 a	19,33 a	2.531,6 a	24,16 a	9,42 a
BR 3302	19,00 a	239,96 a	19,33 a	2.629,2 a	22,83 a	11,14 a
BR 3262	18,83 a	223,13 a	17,50 a	2.198,7 a	22,66 a	12,86 a
BR 3267	19,66 a	250,10 a	19,50 a	2.983,1 a	26,16 a	12,18 a
BR 3299	18,83 a	221,79 a	18,33 a	2.677,1 a	25,19 a	12,52 a
80 kg ha^{-1}	18,83 a	199,24 a	19,83 a	2.320,3 a	23,33 a	5,81 a
50 kg ha^{-1}	15,33 b	183,22 a	20,00 a	2.514,6 a	26,66 a	10,78 a
Control	13,66 b	174,83 a	19,16 a	2.458,7 a	23,83 a	12,42 a
CV (%)	21,91	33,78	13,92	41,30	11,74	45,39

Averages followed by the same letter do not differ according to the Scott-Knott test at 5%.

As the concentration of crude protein reflects the levels of nitrogen in the plants, this result shows that the strains used were efficient in absorbing this element, even being compared to the application of 80 kg ha^{-1} of mineral nitrogen in the fertilizer.

In white oat seeds (*Avena sativa* L.), Kolchinski and Schuch (2003) found increases in the protein content of the kernels as the doses of nitrogen made available to the plants increased.

As was observed at 35 A.D. (Table 3.2), the dry weight of the aerial part and root and the total dry weight at 68 A.D. (Table 3.7) were also not influenced by the treatments, as was the case with the leaf area of the cowpea plants.

Table 3.7 Average values for aerial part dry weight (PSPA), root dry weight (PSR), total dry weight (PST) and leaf area (AF) from cowpea plants harvested at 68 D.A.E., subjected or not to inoculation and nitrogen fertilization.

Treatments	MSPA (g/plant)	MSR (g/plant)	MST (g/plant)	AF (cm^2 /plant)
BR 3301	30,13 a	1,92 a	32,06 a	1.197,0 a
BR 3302	24,26 a	1,57 a	25,83 a	1.020,9 a
BR 3262	31,64 a	1,95 a	33,60 a	1.375,8 a
BR 3267	27,50 a	1,72 a	29,22 a	1.160,7 a
BR 3299	23,54 a	1,49 a	25,03 a	937,8 a
80 kg ha^{-1}	20,52 a	1,84 a	22,37 a	814,7 a
50 kg ha^{-1}	24,92 a	1,61 a	26,52 a	1.075,3 a
Control	22,55 a	1,43 a	23,98 a	891,9 a
CV (%)	47,58	36,10	46,43	50,44

Averages followed by the same letter do not differ according to the Scott-Knott test at 5%.

This shows that the effects of the treatments were not expressed throughout the cycle of this cultivar, even under conditions of regular rainfall, a result that reinforces the data obtained in Chapter I, where there was also no effect of the treatments.

In both cases (Chapters I and III), these responses suggest that the rhizobium-caupi symbiosis was able to fix atmospheric nitrogen and meet the plants' needs, providing similar development to those that received nitrogen fertilization (50 and 80 kg ha).$^{-1}$

However, in some cases (Soares et al., 2006) the vegetative development of the cowpea was influenced by the inoculation of rhizobium strains, and it was found that the INPA 03-11b strain, together with the treatment fertilized with 70 kg of N ha^{-1} , showed the greatest amount of dry weight of the aerial part.

In other crops, such as alfalfa (*Medicago sativa* L.), the production of dry biomass of the aerial part also showed a significant difference between inoculated plants and the control without inoculation with rhizobia (Oliveira et al., 1999).

There were no significant effects of the treatments on the productive characteristics of the cowpea plants (Table 3.8).

Table 3.8 Average values for seed yield (SR), seed yield per plant (RSP), number of pods per plant (NVP), pod weight (PV), pod length (CV) and number of seeds per pod (NSV) of cowpea plants subjected or not to inoculation and nitrogen fertilization.

Treatments	**RS** (kg/ha)	**RSP** (g/plant)	**NVP** (ud/plant)	**BW** (g/plant)	**CV** (cm)	**NSV** (ud/pod)
BR 3301	756,35 a	13,02 a	3,94 a	10,29 a	17,8 a	13,8 a
BR 3302	794,44 a	13,28 a	4,06 a	11,10 a	17,3 a	12,5 a
BR 3262	767,01 a	12,78 a	4,46 a	10,38 a	17,5 a	12,9 a
BR 3267	926,51 a	15,39 a	4,78 a	12,26 a	18,0 a	12,8 a
BR 3299	741,95 a	14,69 a	4,67 a	11,99 a	16,8 a	12,5 a
80 kg ha^{-1}	781,63 a	12,18 a	3,80 a	9,57 a	17,1 a	12,6 a
50 kg ha^{-1}	895,73 a	12,41 a	4,22 a	10,05 a	16,6 a	11,6 a
Control	639,47 a	12,90 a	4,18 a	10,42 a	17,0 a	12,3 a
CV (%)	33,13	39,10	33,49	39,66	6,77	10,06

Averages followed by the same letter do not differ according to the Scott-Knott test at 5%.

The joint analysis of the variables relating to the productive characteristics of the cowpea plants is shown in Table 3.9.

Although there were no differences between the treatments, when comparing some of these variables (RS, RSP, NVP) with those observed in the first planting of the cowpea (Table 3.9), there were big differences, especially in relation to seed yield. This fact reinforces the observation made in Chapter I when it was said that the low rainfall must have been the main reason for the low yields achieved in that first crop of cowpeas.

This variation in productivity can probably also be associated, among other factors, with the population of rhizobial bacteria that nodulate cowpea in the soil. These rhizobia established in the soil are very competitive in forming nodules and probably restricted nodulation by other introduced strains, or were just as efficient.

Table 3.9 Effect of the year of cultivation on the variables seed yield (SR), seed yield per plant (RSP), number of pods per plant (NVP) and pod length (CV) of cowpea plants subjected or not to inoculation and nitrogen fertilization.

TREAT	RS (kg/ha)		RSP (g/plant)		NVP (ud.)		CV (cm)	
	YEAR		YEAR		YEAR		YEAR	
	2005	2006	2005	2006	2005	2006	2005	2006
BR 3301	364,3 b	756,3 a	4,9 b	13,0 a	1,6 b	3,9 a	18,6 a	17,8 a
BR 3302	303,3 b	794,4 a	3,0 b	13,2 a	1,3 b	4,0 a	18,8 a	17,3 a
BR 3262	409,7 a	767,0 a	4,4 a	12,7 a	1,9 a	4,4 a	18,4 a	17,5 a
BR 3267	360,1 b	926,5 a	4,3 b	15,3 a	2,0 b	4,7 a	17,9 a	18,0 a
BR 3299	376,6 b	741,9 a	4,3 b	14,6 a	2,0 b	4,6 a	18,4 a	16,8 b
80 kg ha^{-1}	344,0 b	781,6 a	5,4 b	12,1 a	1,7 b	3,8 a	18,7 a	17,1 b
50 kg ha^{-1}	378,5 b	895,7 a	6,5 b	12,4 a	3,1 a	4,2 a	18,5 a	16,6 b
Control	377,5 b	639,4 a	3,9 b	12,9 a	1,7 b	4,1 a	17,4 a	17,0 a

Averages followed by the same letter in the row do not differ by the F test at 5%.

Another factor that may have had a significant influence on why the inoculated strains did not stand out is the possibility that the cowpea cultivar used is not one of the specific strains used.

Some rhizobium strains (BR 3262, BR 3301, BR 3267, BR 3280, BR 3287, BR 3299) were inoculated into cowpea cultivar BRS Guariba and also showed no significant differences between each other and in relation to the control and fertilized treatments, in all the variables (dry weight of nodules, number of nodules, grain yield) analyzed by Gualter et al. (2006).

The variables that tested the physiological quality of the cowpea seeds are shown in Table 3.10, where there is a significant effect on almost all of them, except for the speed of emergence in sand.

Table 3.10 Average values of seed germination (GER), seedling dry weight (PSP), first count of seedling emergence in sand (PCEA), seedling emergence in sand (EPA) and speed of emergence in sand (IVE) of cowpea submitted or not to inoculation and nitrogen fertilization.

Treatments	GER (%)	PSP (mg/plant)	PCEA (%)	EPA (%)	IVE
BR 3301	87.50 a	54.5 b	78.75 b	86.50 a	17.12 a
BR 3302	89.50 a	57.2 b	82.25 a	85.00 a	25.35 a
BR 3262	79.00 b	51.4 b	71.75 b	74.00 b	14.75 a
BR 3267	82.00 b	52,3 b	75.25 b	78.00 b	31.30 a
BR 3299	87.00 a	50,1 b	87.50 a	87.50 a	17.45 a

80 kg ha^{-1}	96.50 a	71,3 a	86.75 a	94.00 a	18.57 a
50 kg ha^{-1}	79.50 b	50,1 b	76.50 b	78.25 b	15.62 a
Control	91.00 a	50,4 b	86.75 a	90.50 a	18.00 a
CV (%)	7,76	15,06	7,70	7,29	47,41

Averages followed by the same letter do not differ by the Scott-Knott test at 5%.

In terms of seed germination, the highest percentages were obtained when the seeds were produced by plants from nitrogen fertilization (80 kg ha^{-1}), control treatment and seeds inoculated with strains BR 3302, BR 3301 and BR 3299.

As for the dry weight of the seedlings, they were heavier when they came from the treatment that used nitrogen fertilization (80 kg ha^{-1}), an observation that confirms the result obtained for this same variable in Chapter I.

With regard to the first count of emergence in sand, the BR 3299, 80 kg ha^{-1} , control and BR 3302 treatments had the highest percentages (Table 3.10).

In terms of seedling emergence in sand, the best rates were obtained by the treatments 80 kg ha^{-1} , control, BR 3299, BR 3301 and BR 3302.

The data observed for these variables (PCEA, EPA) contrasts with that obtained in the previous planting (Chapter I), since there were no significant differences between the treatments.

Research aimed at analyzing the physiological quality of seeds from plants inoculated with rhizobia is still quite scarce, but there are several studies that have analyzed the influence of nitrogen fertilization on seed quality. These studies can serve as a basis for comparing the results found in this research, since rhizobial inoculation can supply all or part of the nitrogen needs of some crops.

In this regard, Farinelle et al. (2006) evaluated the influence of two soil management systems (conventional and no-till) and nitrogen fertilization (0, 40, 80, 120 and 160 kg ha^{-1}) on the productivity and physiological quality of bean seeds (*Phaseolus vulgaris* L.). The results found by these authors indicated that both the yield and the physiological quality (germination and vigor) of these seeds increased as the doses of mineral nitrogen in the top dressing increased.

Positive results from increasing nitrogen doses were also observed in beans, when the germination, IVG and field emergence of these seeds were evaluated (Oliveira et al., 2003).

On the other hand, in some studies, the supply of mineral nitrogen to the bean crop did not provide significant differences in the physiological quality of its seeds (Ambrosano et al.,

1999; Crusciol et al., 2003; Meira et al., 2005).

Table 11 shows the joint analysis of the experiments carried out in the two consecutive years of cowpea planting (2005 and 2006), involving the variables relating to the physiological quality of the seeds.

For most of the treatments, it can be seen that the variables showed greater responses when the seeds were produced in the first year of planting. One of the main reasons for this result may have been the fact that the seeds came from different batches and/or agricultural harvests, so they may have been produced under different conditions and therefore showed differences in germination percentage in the different years of planting.

Table 11. Effect of the agricultural year on seed germination (GER), seedling dry weight (PSP), first count of seedling emergence in sand (PCEA) and seedling emergence in sand (EPA) of cowpea submitted or not to inoculation and nitrogen fertilization.

TREAT	**GER** (%)		**MSP** (mg/plant)		**PCEA** (%)		**EPA** (%)	
	YEAR		YEAR		YEAR		YEAR	
	2005	2006	2005	2006	2005	2006	2005	2006
BR 3301	98,0 a	87.5 b	55,5 a	54.5 a	89,0 a	78.7 b	97,0 a	86.5 b
BR 3302	100,0 a	89.5 b	57,5 a	57.2 a	92,0 a	82.2 b	95,0 a	85.0 b
BR 3262	99,0 a	79.0 b	52,3 a	51.4 a	90,0 a	71.7 b	93,0 a	74.0 b
BR 3267	95,5 a	82.0 b	54,3 a	52,3 a	88,0 a	75.2 b	92,0 a	78.0 b
BR 3299	94,5 a	87.0 b	52,9 a	50,1 a	95,0 a	87.5 a	95,0 a	87.5 a
80 kg ha^{-1}	95,5 a	96.5 a	72,7 a	71,3 a	86,0 a	86.7 a	93,0 a	94.0 a
50 kg ha^{-1}	93,0 a	79.5 b	50,9 a	50,1 a	90,0 a	76.5 b	92,0 a	78.2 b
Control	97,5 a	91.0 a	52,9 a	50,4 a	93,0 a	86.7 a	97,0 a	90.5 a

Averages followed by the same letter in the row do not differ by the F test at 5%.

4.4 Conclusions

The strains that have not yet been officially recommended (BR 3262 and BR 3299) for cowpeas showed similar efficiency to the other strains and to the treatments fertilized (80 and/or 50 kg ha^{-1}) with mineral nitrogen, when evaluating seed production;

Plants inoculated with strains BR 3301, BR 3302 and BR 3299 produced seeds with the highest germination rates, which did not differ from the 80 kg ha^{-1} and control treatments;

The most vigorous seeds came from plants that received 80 kg of N ha ;$^{-1}$

The native strains (control) performed similarly to those inoculated in the two years of cowpea cultivation.

4.5 Bibliographical references

ALCÂNTARA, R. M. C. M. de; FORTALEZA, J. M.; XAVIER, G. R. e SOUZA, J. S. de. Inoculation of cowpea [Vigna Unguiculata (L.) Walp.] with rhizobium BR 3267 in Teresina, Pi. In: CONGRESSO NACIONAL DE FEIJÂO-CAUPI, 1., 2006, Teresina. **Proceedings**... Teresina: Embrapa, 2006. Available at < http://www.cpamn.embrapa.br/anaisconac2006> Accessed on January 26, 2007.

AMBROSANO, E.J.; AMBROSANO, G.M.B.; WUTKE, E.B.; BULISANI, E.A.; MARTINS, A.L.M.; SILVEIRA, L.C.P. Effects of nitrogen and micronutrient fertilization on seed quality of the bean cultivar iac-carioca. **Bragantia**, Campinas, v. 58, n.2, p.393-399, 1999.

BRAZIL. Ministry of Agriculture and Agrarian Reform. **Rules for Seed Analysis**. Brasilia: SNDA/DNDV/CLAV, 1992. 365p.

CRUSCIOL, C.A.C.; EDUARDO LIMA, E.V.; ANDREOTTI, M.; NAKAGAWA, J.; LEMOS, L.B.;, MARUBAYASHI, O.M. Effect of nitrogen on the physiological quality, productivity and characteristics of bean seeds. **Revista Brasileira de Sementes**, Brasilia, v. 25, n. 1, p.108-115, 2003.

FARINELLI, R.; LEMOS, L.B.; CAVARIANI, C.; NAKAGAWA, J. Productivity and physiological quality of bean seeds as a function of soil management systems and nitrogen fertilization. **Revista Brasileira de Sementes**, Brasilia, v. 28, n. 2, p.102-109, 2006.

FRANCO, A.A.; DOBEREINER, J. Soil biology and the sustainability of tropical soils. **Summa Phitopatologica**, Jaboticabal, v. 20, n. 1, p. 68-74, 1994.

GUALTER, R.M.R.; HENRIQUES NETO, D.; LEITE, L.F.C.; DANTAS, J.S.; SOUSA, F.P. Inoculation of *Bradyrhizobium* spp. strains and their effects on nodulation and productivity of cowpea in the Piauí highlands. In: CONGRESSO NACIONAL DE FEIJÂO-CAUPI, 1., 2006, Teresina. **Proceedings**... Teresina: Embrapa, 2006. Available at < http://www.cpamn.embrapa.br/anaisconac2006> Accessed on January 26, 2007.

KOLCHINSKI, E.M.; SCHUCH, L.O.B. Attributes of industrial performance and seed quality in white oats as a function of nitrogen fertilizer availability. **Ciência Rural**, Santa Maria, v.33, n.3, p.587-589, 2003.

LIMA, A.S.; PEREIRA, J.P.A.R.; MOREIRA, F.M.S. Phenotypic diversity and symbiotic efficiency of *Bradyrhizobium* spp. strains from Amazonian soils. **Pesquisa Agropecuâria Brasileira**, Brasilia, v.40, n.11, p.1095-1104, nov. 2005.

MAGUIRE, J.D. Speed of germination-aid in selection and evaluation for seedling emergence and vigor. **Crop Science**, Madison, v.2, n.1, p.176-177,1962.

MALAVOLTA, E.; VITTI, G.C.; OLIVEIRA, S.A. de. **Evaluation of plant nutritional status - principles and applications**. Piracicaba: POTAFOS, 1989. 201p.

MARTINS, L. M.; XAVIER, G. R.; RANGEL, F. W.; RIBEIRO; J. R. A.; NEVES, M. C. P.; MORGADO, L. B.; RUMJANEK, N. G. Contribution of biological nitrogen fixation to cowpea: a strategy for improving grain yield in the semi-arid region of Brazil. **Biology and Fertility of Soils**, v.38, p.333-339, 2003.

MEDEIROS, R.D.; ARAÙJO, W.F.; COSTA, M.C. Efeito de sistemas de preparo do solo e métodos de irrigaçâo sobre a cultura do caupi em vàrzeas em Roraima. **Revista Brasileira de Engenharia Agricola e Ambiental**, Campina Grande, v.9, n.2, p.205-209, 2005.

MEIRA, F.A.; SA, M.E.; BUZETTI, S.; ARF, O. Doses and times of nitrogen application to irrigated no-till bean crops. **Pesquisa Agropecuâria Brasileira**, Brasilia, v.40, n.4, p.383-388, 2005.

MORGADO, L.B.; MARTINS, L.M.V.; XAVIER, G.R. and RUMJANEK, N.G. Evaluation of the potential of rhizobium strains to fix nitrogen associated with cowpea in Petrolina-PE. In: CONGRESSO NACIONAL DE FEIJÂO-CAUPI, 1., 2006, Teresina. **Proceedings**... Teresina: Embrapa, 2006. Available at < http://www.cpamn.embrapa.br/anaisconac2006> Accessed on January 26, 2007.

MPEPEREKI, S; WOLLUM, A.G.; MAKONESE, F. Diversity in symbiotic specificity of cowpea indigenous to Zimbabwean soils. **Plant and Soil**. Dordrecht, v. 186, p. 167-171, 1996.

NAKAGAWA, J. Vigor tests based on seedling performance. In: KRZYZANOWSKI, F.C., VIEIRA, R.D., FRANÇA NETO, J.B. **Seed vigor**: concept and tests. Londrina: ABRATES. 1999. 218p.

OLIVEIRA, A.P.; PEREIRA, E.L.;, BRUNO, R.L.A.; ALVES, E.U.; COSTA, R.F.; LEAL, F.R.LF. Yield and physiological quality of cowpea seeds as a function of nitrogen sources and doses. **Revista Brasileira de Sementes**, Brasilia, v. 25, n. 1, p.49-55, 2003.

OLIVEIRA, P.P.A.; TSAI, S.M.; CORSI, M.; DiAZ, M.D.P. Interaction between cultivars, commercial rhizobium meliloti strains and fungicides in increasing alfalfa production. **Pesquisa Agropecuâria Brasileira**, Brasilia, v.34, n.3, p.425-431, mar. 1999.

PEDROSA, F.O.; HUNGRIA, M.; YATES, G.; NEWTON, W.E. **Nitrogen Fixation:** from molecules to crop productivity. Dordrecht: Kluwer Academic Publishers. 2000. 700p.

PEREIRA, K,C. **Morphological, biochemical and molecular characterization of rhizobia recommended for inoculation of tree legumes**. 2002, 63 f.

Dissertation (Master's Degree in Microbiology) - Faculty of Agricultural and Veterinary Sciences, Universidade Estadual Paulista, Jaboticabal, 2002.

SANTOS, C.A.F.; ARAÙJO, F.P.; MENEZES, E.A. Productive behavior of caupi in irrigated and rainfed regimes in Petrolina and Juazeiro. **Pesquisa Agropecuâria Brasileira**, Brasilia, v.35, n.11, p.2229-2234, 2000.

SILVA, V.N.; SILVA, L.E.S.F.; FIGUEIREDO, M.V.B. Atuação de rizóbios com rizobactéria promotora de crescimento em plantas na cultura do caupi (*Vigna unguiculata* [L.] Walp.). **Acta Scientiarum Agronomy**. Maringà, v. 28, n. 3, p. 407412, 2006.

SINGLETON, P.W.; BOHLOOL, B.B.; NAKAO, P.L. Legume response to rhizobial inoculation in the tropics: myths and realities. In: LAL, R.; SANCHEZ, P.A. (Ed.). **Myths and science of soils of the tropics**. Madison: Soil Science Society of American/American Society of Agronomy, 1992. p. 135-155. (SSSA Special Publication, 29).

SOARES, A.L.; PEREIRA, J.P.A.R.; FERREIRA, P.A.A.F.; VALE, H.M.M.; LIMA, A.S.; ANDRADE, M.J.B.; MOREIRA, M.S.. Agronomic efficiency of selected rhizobia and diversity of native nodulating populations in Perdoes (MG). I - caupi. **Revista Brasileira de Ciência do Solo**, Viçosa, v. 30, p.795-802, 2006.

TEIXEIRA, S.M.; MAY, P.H.; SANTANA, A.C. de. Production and economic importance of caupi in Brazil. In: ARAUJO, J.P.P. de; WATT, E.E. (Org.) **O caupi no Brasil**. Brasilia: IITA/Embrapa, 1988. p.99-136.

YOKOYAMA, L.P.; WETZEL, C.T.; VIEIRA, E.H.N.; PEREIRA, G.V. **Bean seeds: production, use and marketing**. In: Vieira, E.H.N.; Rava, C.A. In: Bean seeds: production and technology. Santo Antônio de Goiàs: EMBRAPA Arroz e feijâo, 2000. chap.12, p.249-270.

ZILLI, J.E.; VALICHESKIR.R.; RUMJANEK, N.G.; SIMÔES-ARAÙJO, J.L.; FREIRE FILHO, F.R. E NEVES, M.C.P. Symbiotic efficiency of *Bradyrhizobium* strains isolated from Cerrado soil on caupi. **Pesquisa Agropecuâria Brasileira**, Brasilia, v.41, n.5, p.811-818, 2006.

APPENDIX

Summary of the analysis of variance of the variables studied in Chapter I.

Sources of Variation	GL	NN	PSN	PN	EfR	PSPA (35)	PSPA (55)	PSPA (75)	PSR (35)	PSR (55)	PSR (75)
		Mean Squares									
Treatments	7	52,12	2176,76*	3,63**	913,78	0,11	0,58	11,05	0,006	0,05	0,04
Block	3	14,20	343,07	1,11	16259,36	0,57	1,64	4,95	0,01	0,04	0,01
Waste	21	33,39	653,65	1,04	1139,72	0,23	0,89	12,86	0,007	0,05	0,03
CV		35,70	40,41	26,15	29,64	41,91	32,52	47,74	26,59	27,84	26,28
		CS	LS	ES	PS	RS	RSP	NSP	NVP	NSV	
Treatments	7	0,06	0,18**	0,16*	1,49**	3849,11	4,46	75,42	1,16	1,35	
Block	3	0,17	0,16	0,09	0,60	52892,25	32,86	518,61	3,17	3,91	
Waste	21	0,22	0,05	0,06	0,28	17146,27	5,32	90,04	0,91	1,44	
CV		5,29	3,12	4,61	2,19	35,94	49,79	49,70	49,07	9,90	
		GER	MSP	PCEA	EPA	IVE					
Treatments	7	22,78*	130,14*	33,07	16,85	0,64					
Waste	24	7,30	78,92	43,35	13,80	0,56					
CV		2,80	16,00	7,29	3,94	4,03					

Key: Number of nodules (NN); nodule dry weight (PSN); nodule weight (PN); relative plot efficiency (EfR); aerial part dry weight (PSPA 35, 55 and 75 D.A.E.); root dry weight (PSR 35, 55 and 75 D.A.E.); seed length (CS); seed width (LS); seed thickness (ES); weight of 1. 000 seeds (PS); seed yield (RS); seed yield per plant (RSP); number of pods per plant (NVP); seed length (CS); seed width (LS); seed thickness (ES).000 seeds (PS); seed yield (RS); seed yield per plant (RSP); number of seeds per plant (NSP); number of pods per plant (NVP); pod length (CV); number of seeds per pod (NSV); seed germination (GER); seedling dry weight (MSP); first count of seedling emergence in sand (PCEA); seedling emergence in sand (EPA) and speed of emergence in sand (IVE).

Summary of the analysis of variance of the variables studied in Chapter II.

Sources of variation	GL	Mean Squares									
		PNA	CNA	PNS	CNS	PNN	CNN	PPA	CPA	PPS	CPS
Treatments	7	0,12	285,57	0,10	8592,37	1,77**	8,11**	0,02	17,61	0,008	427,11
Block	3	0,49	1883,20	0,11	43080,46	0,70	2,17	0,08	136,50	0,01	3288,89
Waste	21	0,13	519,54	0,07	8160,27	0,35	2,15	0,03	32,25	0,01	556,69
CV		8,87	46,72	7,19	51,82	11,36	43,27	14,89	41,26	10,64	48,80
		PPN	CPN	PKA	CKA	PKS	CKS	PKN	CKN		
Treatments	7	0,03**	0,26*	1,11**	245,75	0,004	1251,03	0,05	0,62*		
Block	3	0,04	0,08	0,79	1291,06	0,34	16059,59	0,44	0,56		
Waste	21	0,008	0,07	0,27	297,50	0,01	1536,17	0,06	0,21		
CV		9,39	43,08	18,76	52,29	8,97	51,92	15,17	44,92		

Caption: Percentage of nitrogen in the aerial part (PNA); nitrogen content in the aerial part (CNA); percentage of nitrogen in the seed (PNS); nitrogen content in the seed (CNS); percentage of nitrogen in the nodule (PNN) and percentage of nitrogen in the nodule (CNN); percentage of phosphorus in the aerial part (PPA); percentage of phosphorus in the aerial part (CPA); percentage of phosphorus in the seed (PPS); phosphorus content in the seed (CPS); percentage of phosphorus in the nodule (PPN) and percentage of phosphorus in the nodule (CPN); percentage of potassium in the aerial part (PKA); potassium content in the aerial part (CKA); potassium percentage in the seed (PKS); potassium content in the seed (CKS); potassium percentage in the nodule (PKN) and potassium content in the nodule (CKN).

A 3. Summary of the analysis of variance of the variables studied in Chapter III.

Sources of Variation	**GL**	**Mean Squares**									
		NuN	MSN	MSA	MSR	MST	PNA	CNA	PNS	CNS	PNN
Treatments	7	52,44	442,05	0,08	0,01	0,13	0,68	130,75	0,09	8308,76	0,19
Block	3	40,20	642,79	0,39	0,02	0,59	0,32	277,43	0,07	13403,29	0,13
Waste	21	24,34	408,96	0,09	0,008	0,14	0,38	128,69	0,15	27790,65	0,21
CV		32,10	45,08	25,75	24,35	24,51	21,81	33,92	12,75	41,19	11,95
		CNN	PPBA	CPBA	PPBS	CPBS	PPBN	CPBN	PSPA	PSR	PST
Treatments	7	0,84	29,13*	4756,38	4,08	338712,83	13,54	33,07	87,58	0,24	93,63
Block	3	1,61	8,38	11564,49	4,95	563404,83	7,22	63,10	128,57	0,43	140,76
Waste	21	0,62	15,48	5011,47	7,08	1099898,98	8,17	24,45	148,80	0,37	161,02
CV		45,41	21,91	33,78	13,92	41,30	11,74	45,39	47,58	36,10	46,43
		AF	RS	RSP	NVP	PV	CV	NSV			
Treatments	7	200366,19	48425,67	7,60	0,72	5,39	1,32	2,27			
Block	3	101617,01	124138,93	21,43	0,69	8,21	2,93	1,58			
Waste	21	285559,62	68129,82	27,19	2,04	18,21	1,37	1,62			
CV		50,44	33,13	39,10	33,49	39,66	6,77	10,06			
		GER	MSP	PCEA	EPA	IVE					
Treatments	7	146,28	200,13	144,62	188,81	127,65					
Waste	24	45,00	70,77	38,62	37,67	87,86					
CV		7,76	15,06	7,70	7,29	47,41					

Legend: Number of nodules (NuN); nodule dry weight (MSN); aerial dry weight (MSA); root dry weight (MSR) and total dry weight (MST); aerial nitrogen percentage (PNA); aerial nitrogen content (CNA); seed nitrogen percentage (PNS); seed nitrogen content (CNS); percentage of nitrogen in the nodule (PNN) and content of nitrogen in the nodule (CNN); percentage of crude protein in the aerial part (PPBA); content of crude protein in the aerial part (CPBA); percentage of crude protein in the seed (PPBS); content of total crude protein in the seed (CPBS); percentage of crude protein in the nodule (PPBN) and crude protein content in the nodule (CPBN); aerial part dry weight (PSPA); root dry weight (PSR); total dry weight (PST) and leaf area (AF); seed yield (RS); seed yield per plant (RSP); number of pods per plant (NVP); pod weight (PV); pod length (CV) and number of seeds per pod (NSV); seed germination (GER); seedling dry weight (PSP); first count of seedling emergence in sand (PCEA); seedling emergence in sand (EPA) and speed of emergence in sand (IVE).

Printed by Books on Demand GmbH, Norderstedt / Germany